Time-Frequency Transforms for Radar Imaging and Signal Analysis

For a listing of recent titles in the *Artech House Radar Library*, turn to the back of this book.

Time-Frequency Transforms for Radar Imaging and Signal Analysis

Victor C. Chen
Hao Ling

Artech House
Boston • London
www.artechhouse.com

Library of Congress Cataloging-in-Publication Data
Chen, Victor C..
  Time-frequency transforms for radar imaging and signal analysis / Victor C. Chen,
  Hao Ling.
    p. cm. — (Artech House radar library)
  Includes bibliographical references and index.
  ISBN 1-58053-288-8 (alk. paper)
    1. Radar—Mathematics. 2. Signal processing—Mathematics. 3. Imaging
  systems—Mathematics. 4. Time-domain analysis. I. Ling, Hao. II. Title.
  III. Series.
TK6578.C44 2002
621.3848—dc21   2001055229

British Library Cataloguing in Publication Data
Chen, Victor C.
  Time-frequency transforms for radar imaging and signal analysis. — (Artech House
  radar library)
  1. Radar 2. Signal processing 3. Time-series analysis 4. Frequency spectra
  I. Title II. Ling, Hao
  621.3'848

  ISBN 1-58053-288-8

Cover design by Igor Valdman

International Standard Book Number: 1-58053-288-8
Library of Congress Catalog Card Number: 2001055229

10 9 8 7 6 5 4 3 2 1

Contents

Foreword   *xi*

Preface   *xiii*

Acknowledgments   *xvii*

1   Introduction   **1**

1.1   Electromagnetic Back-Scattering from Targets   6

1.2   Radar Signal and Noise   9
1.2.1 Signal Waveforms   9
1.2.2 The SNR   11

1.3   Radar Ambiguity Function and Matched
       Filter   13
1.3.1 Radar Ambiguity Function   13
1.3.2 Matched Filter   17
1.3.3 Pulse Compression   19

1.4   Synthetic Aperture Radar Imaging   20
1.4.1 Range Profile   21
1.4.2 Range Resolution   21

1.4.3	Cross-Range Resolution	22
	References	22
2	**Time-Frequency Transforms**	**25**
2.1	Linear Time-Frequency Transforms	26
2.1.1	The STFT	28
2.1.2	The CWT	32
2.1.3	Adaptive Time-Frequency Representation	33
2.2	Bilinear Time-Frequency Transforms	36
2.2.1	The WVD	37
2.2.2	Cohen's Class	39
2.2.3	The TFDS	42
	References	45
3	**Detection and Extraction of Signal in Noise**	**47**
3.1	Introduction	48
3.2	Time-Varying Frequency Filtering	48
3.3	SNR Improvement in the Time-Frequency Domain	51
3.3.1	SNR Definition Suitable for Signal Detection and Extraction	54
3.3.2	SNR in the Joint Time-Frequency Domain	56
3.4	CFAR Detection in the Joint Time-Frequency Domain	57
3.5	Signal Extraction in the Joint Time-Frequency Domain	61
3.5.1	Time-Frequency Expansion and Reconstruction	61
3.5.2	Time-Frequency Masking and Signal Extraction	62
	References	63

4	**Time-Frequency Analysis of Radar Range Profiles**	**65**
4.1	Electromagnetic Phenomenology Embedded in Back-Scattered Data	66
4.2	Time-Frequency Representation of Range Profiles	70
4.3	Application of High-Resolution Time-Frequency Techniques to Scattering Data	77
4.3.1	Use of the CWT	77
4.3.2	Use of the TFDS	79
4.3.3	Windowed Superresolution Algorithm	81
4.3.4	Adaptive Gaussian Representation	83
4.4	Extraction of Dispersive Scattering Features from Radar Imagery Using Time-Frequency Processing	85
	References	89
5	**Time-Frequency-Based Radar Image Formation**	**93**
5.1	Radar Imaging of Moving Targets	94
5.2	Standard Motion Compensation and Fourier-Based Image Formation	102
5.3	Time-Frequency-Based Image Formation	104
5.4	Radar Imaging of Maneuvering Targets	107
5.4.1	Dynamics of Maneuvering Targets	107
5.4.2	Radar Imaging of Maneuvering Target Using Time-Frequency-Based Image Formation	108
5.5	Radar Imaging of Multiple Targets	113
5.5.1	Multiple-Target Resolution Analysis	113
5.5.2	Time-Frequency-Based Phase Compensation for Multiple Targets	117

5.5.3	Time-Frequency-Based Image Formation for Radar Imaging of Multiple Targets	119
5.6	Summary	120
	References	120

6 **Motion Compensation in ISAR Imaging Using Time-Frequency Techniques** **123**

6.1	Motion Compensation Algorithms	124
6.2	Time-Frequency-Based Motion Compensation	126
6.2.1	Estimating Phase Using Adaptive Time-Frequency Projection	128
6.2.2	Motion Error Elimination	129
6.3	Motion Compensation Examples of Simulated and Measured Data	131
6.4	Presence of 3D Target Motion	135
	References	144

7 **SAR Imaging of Moving Targets** **147**

7.1	Radar Returns of Moving Targets	148
7.1.1	Range Curvature	149
7.1.2	Clutter Bandwidth	150
7.1.3	Analysis of Radar Returns from Moving Targets	152
7.2	The Effect of Target Motion on SAR Imaging	155
7.3	Detection and Imaging of Moving Targets	157
7.3.1	Single-Aperture Antenna SAR	157
7.3.2	Multiple-Antenna SAR	161
7.4	SAR Imaging of Moving Targets Using Time-Frequency Transforms	165
7.4.1	Estimation of Doppler Parameters Using Time-Frequency Transforms	166

7.4.2   Time-Frequency-Based SAR Image Formation
        for Detection of Moving Targets   168
        References   170

**8   Time-Frequency Analysis of Micro-Doppler
        Phenomenon   173**

8.1   Vibration-Induced Micro-Doppler   174
8.1.1   Time-Frequency Signature of a Vibrating
        Scatterer   177
8.1.2   An Example of Micro-Doppler Signatures of
        Moving Targets   179

8.2   Rotation-Induced Micro-Doppler   181
8.2.1   Rotor Blade Motion   181
8.2.2   Radar Returns from Rotor Blades   181
8.2.3   Time-Domain Signatures of Rotation-Induced
        Modulations   184
8.2.4   Frequency-Domain Signatures   188
8.2.5   Time-Frequency Signatures   190
        References   192

**9   Trends in Time-Frequency Transforms for
        Radar Applications   193**

9.1   Applications of Adaptive Time-Frequency
        Transforms   193

9.2   Back-Scattering Feature Extraction   194

9.3   Image Formation   195

9.4   Motion Compensation   195

9.5   Moving Target Detection   196

9.6   Micro-Doppler Analysis   198
        References   199

List of Acronyms   203

About the Authors   205

Index   207

Foreword

The exposition and exploitation of joint time-frequency methods of signal analysis is at present an area of vigorous research and development, and many international conferences and symposia address this topic.

The mathematical basis for joint time-frequency signal analysis has been established and theoretically understood for quite a few years; however, the computational requirements for real-time signal processing and effective graphical visualization of results has exceeded commonly available computers until quite recently. Today, affordable workstations or personal computers with sufficient computing and graphical display capability are readily available to implement real-time time-frequency transforms and dynamically visualize time-dependent aspects of spectral signal structure. These are exactly the tools necessary to exploit joint time-frequency methods for radar-signal analysis and imaging of moving targets.

Drs. Chen and Ling are each forging new ways of applying time-frequency processing to radar-signal analysis, radar imaging, and extraction of target features from moving targets. They combine their research efforts in this book to produce the first self-contained description of joint time-frequency processing methods uniquely adapted to radar applications.

A concise review of radar and time-frequency transforms is provided as background needed to appreciate how joint time-frequency processing methods can improve conventional time or frequency processing methods. The book then describes and illustrates the advantages of using joint time-frequency processing for radar signal detection, range profile analysis, synthetic aperture radar, inverse synthetic aperture radar imaging, and micro-

Doppler signal analysis. The last chapter describes current trends regarding time-frequency transforms for radar applications and indicates on-going research topics. In summary, the contents of this book are well balanced between pedagogical and current research material, and the many illustrations facilitate comprehension of the material presented.

The ultimate goal of radar-signal analysis and radar imaging is not merely to detect or form a picture of the target, but to automate processing that contributes to the identification of the target. For noncooperative targets, this goal remains elusive. This timely book presents new tools that address this problem, and it is hoped that it will stimulate even more ideas.

William J. Miceli
Associate Director
The Office of Naval Research
International Field Office

Preface

Joint time-frequency analysis has been a topic of much interest in the signal processing community in the past decade. The vigorousness of research activities is especially evident from the number of conferences and special topic sessions dedicated to joint time-frequency representations. Over the past ten years, time-frequency transforms have also been investigated by radar researchers as a unique tool for radar-specific signal analysis and image processing applications. Both traditional time-frequency techniques, as well as the new tools developed in the signal processing community, have been applied to various radar problems. Like the developments in other fields, such as underwater acoustics and speech processing, it was found that time-frequency transforms provide additional insight into the analysis, interpretation, and processing of radar signals that is sometimes superior to what is achievable in the traditional time or frequency domain alone. The specific applications where time-frequency transforms have been used include signature analysis and feature extraction, motion compensation and image formation, signal denoising, and imaging of moving targets.

The intent of this book is to provide a summary of the authors' research into applying time-frequency transforms to radar applications. In particular, our focus is on the extraction of target features from the radar-backscattered signal for the purpose of signature diagnostics and automatic target recognition. Both one-dimensional range profiles and two-dimensional radar imagery, two traditional feature spaces for mapping the geometrical details of the target, are considered. We describe time-frequency techniques for extracting other structural features on the target due to higher-order scattering mechanisms and complex target motions. Our objective is to document the progress made in this area of research and provide a handy reference for researchers interested in this field. In the process, we hope that this book will stimulate additional work in this area and lead to further advances in the state of the art.

This book is organized as follows: In Chapter 1, we provide a brief introduction to basic radar concepts, including radar-backscattering principles, radar waveforms, noise and clutter, ambiguity functions, match filter, and pulse compression. We also introduce the idea of radar imaging, including both synthetic aperture radar (SAR) and inverse synthetic aperture radar (ISAR). In Chapter 2, we provide an overview of joint time-frequency transforms. Our discussions cover both linear transforms, such as the short-time Fourier transform, the continuous wavelet transform and the adaptive spectrogram, and bilinear transforms, including the Wigner-Ville distribution, Cohen's class, and the time-frequency distribution series. In Chapter 3, we discuss the use of time-frequency transforms in the detection and extraction of radar signal in noise. The concept of the time-varying frequency filter is introduced for the denoising of radar signal in noise.

Chapters 4 and 5 discuss the main developments of time-frequency transforms for radar-signal analysis and image processing. In Chapter 4, we describe the use of time-frequency transforms for one-dimensional radar range profiles. In particular, we focus on how complex electromagnetic scattering mechanisms can be better analyzed and interpreted in the joint time-frequency space. We also illustrate the use of high-resolution time-frequency transforms for localizing and extracting the time-frequency scattering features. In Chapter 5, the use of time-frequency transforms for two-dimensional radar-image formation is detailed. We discuss radar imaging of moving targets and the time-varying behavior of their Doppler shifts, as well as how time-frequency analysis is used for radar imaging of multiple targets.

Chapters 6, 7, and 8 describe more specific applications of time-frequency transforms for radar. In Chapter 6, we discuss the use of joint time-frequency analysis for ISAR motion compensation. In Chapter 7, we discuss the use of time-frequency transforms in SAR imaging of moving targets. In Chapter 8, we describe the use of time-frequency transforms to analyze the micro-Doppler phenomenon on targets. Finally, in Chapter 9, we conclude by providing current trends and future outlooks in applying time-frequency transforms for radar applications. We should point out that this book mainly emphasizes the application perspective (i.e., how and what type of time-frequency transform can be used to carry out radar signal processing and extract relevant target information effectively). Although we give an overview of the time-frequency tools, no detailed theoretical treatment is provided. For more fundamental discussion of time-frequency analysis, the readers are referred to the two excellent books on joint time-frequency analysis, one by Leon Cohen and one by Shie Qian and Dapang Chen (Chapter 2, [2, 3]).

Acknowledgments

This book would not have been possible without the generous support and help of a number of people. First and foremost, we would like to thank William Miceli of the Office of Naval Research, who has provided technical leadership and research funding for a significant portion of the work reported in this book. We are also grateful to him for his stimulating discussions over the years and his encouragement in preparing this book. Dr. Ling would also like to thank the Department of Defense Joint Services Electronics Program, which funded his joint time-frequency research from 1992 to 1998, for its support. We also gratefully acknowledge the partial support of the Air Force Wright Laboratory and the Air Force Office of Scientific Research MURI Program.

Dr. Ling would like to express his sincere thanks to his former and current students, Hyeongdong Kim, Yuanxun Wang, Luiz Trintinalia, Rajan Bhalla, John Moore, Junfei Li, Caner Ozdemir, Hai Deng, and Tao Su, who conceived and carried out most of the research detailed in this book.

The authors would also like to thank many of their mentors, colleagues, and friends—Shung-Wu "Andy" Lee, Dennis Andersh, Francis Bostick, Edward Powers, Charles Liang, Xiang-Gen Xia, Jian Li, Ben Flores, Tatsuo Itoh, Weng Chew, and Yuen-Tze Lo—for their technical advice, moral support, and warm friendship. Special thanks go to Shie Qian, who has worked with each of us and should be credited with initiating us into the exciting field of joint time-frequency analysis.

Finally, the authors would like to express their warmest appreciation to their families. Without their tremendous love and support, this book would not have been possible.

1

Introduction

Radar is an electromagnetic instrument used for the detection and location of targets, such as aircraft, ships, and ground vehicles. It transmits electromagnetic energy to a target and receives the reflected signal from the target and clutter as illustrated in Figure 1.1. Any unwanted radar return that can interfere with the detection of the desired targets is referred to as clutter. From the received radar signal, target-related information such as location and velocity can be accurately measured. Compared to optical and infrared sensors, the radar as a radio frequency (RF) sensor can perform at long range, with high accuracy, and under all weather conditions. Therefore, it has been widely used for civilian and military purposes [1–3].

Suppose a radar transmits a signal $s_T(t)$ at RF f_0. The received signal $s_R(t)$ reflected from a target is proportional to the transmitted signal with a round-trip delay $s_T(t - \tau)$ and scaled by the reflectivity function ρ of the target,

$$s_R(t) \propto \rho s_T(t - \tau) = \rho \exp\{j\omega_0(t - \tau)\} \qquad (1.1)$$
$$= \rho \exp\{j2\pi f_0(t - \tau)\} \quad (0 \le t \le T)$$

where T is the time duration of the signal and $\omega_0 = 2\pi f_0$ is the angular frequency of the signal.

The round-trip travel time τ is determined by

$$\tau = \frac{2R}{c} \qquad (1.2)$$

1

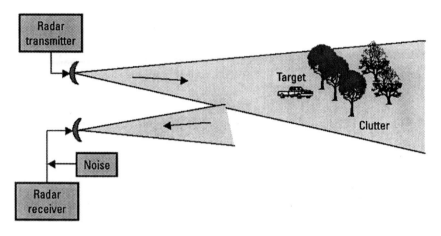

Figure 1.1 Radar operational scenario.

where R is the range from the radar to the target and c is the speed of electromagnetic wave propagation.

When the target is moving with a velocity V_R relative to the radar, called the radial velocity, the radar signal must travel a longer or shorter distance to reach the target. The signal received at time t is reflected from the target at time $(t - \tau(t)/2)$, and the round-trip travel time is a time-varying delay $\tau(t)$.

In addition to the signal reflected from the target, there is also additive noise. The signal-to-noise ratio (SNR) at the radar receiver is determined by the intensity of the received signal, the noise figure, and bandwidth of the receiver. Any improvement in SNR will increase the probability of the target detection and the accuracy of parameter estimation.

Radar usually transmits a sequence of pulses or other signal waveforms at a pulse repetition frequency (PRF) required by the maximum range of detection. In the radar receiver, the received RF signal is first converted to an intermediate frequency (IF) signal. Then, the IF signal is converted into two video frequency signals, the in-phase and the quadrature-phase (I and Q) components, using two synchronous detectors that have an identical reference signal but 90-degree phase difference between them. The I and Q signals can preserve the phase information contained in the IF signal and, thus, enable the positive and the negative Doppler frequency shift to be distinguished [1–3].

Target information embedded in the returned signals may be examined directly from the radar range profile [i.e., the distribution of target reflectivity along the radar line of sight (LOS) to the target] or from its frequency spectrum by applying the Fourier transform [4–7]. The target's range measured along the radar LOS can be estimated by the time-delay between the transmitted signal and the received signal. For a moving target, its velocity is measured based on the well-known Doppler effect. If the radar transmits a signal at a frequency f_0, the reflected signal from the moving target is subjected to a Doppler frequency shift f_D from its transmitted frequency f_0 induced by the relative motion between the radar and the target. In the case where a target has a radial velocity V_R, the Doppler frequency shift f_D is determined by the radial velocity of the target and the radar transmitted frequency f_0:

$$f_D = -2f_0 \frac{V_R}{c} \tag{1.3}$$

where V_R is defined as a positive value when the target is moving away from the radar. Therefore, if a target is moving towards the radar at a velocity $V_R = -1000$ (ft/s) $= -304.8$ (m/s), the Doppler frequency shift for X-band radar operated at 9,842 MHz is +20 kHz.

Radar targets, especially man-made targets, can be considered as a collection of point-scatterers. These scatterers may have a large variety of reflecting or back-scattering behaviors [8, 9]. They can be surfaces, edges, corners, dihedrals, trihedrals, and cavities (Figure 1.2). Each type of scatterer has a different back-scattering behavior.

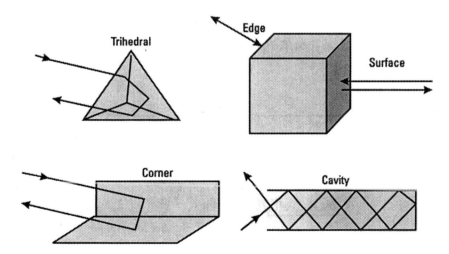

Figure 1.2 Man-made targets: surface, edge, corner, trihedral, and cavity.

Radar range profiles can provide target information about target length and positions of strong scatterers, such as a radar dish, engine intakes, and other scattering centers as shown in Figure 1.3. The dimension transverse to the radar LOS is called the cross range. Because the Doppler shift of a scatterer on a target is proportional to the cross range of the scatterer by a scaling factor, the projection of the target reflectivity distribution on the cross-range dimension can be obtained from the distribution of Doppler shifts, called the Doppler profile. With high-resolution Doppler profiles, the locations of strong scatterers and the target's extents in the Doppler dimension can also be obtained as illustrated in Figure 1.4. By combining range profiles and Doppler profiles, a two-dimensional (2D) radar image may be generated [9–12]. A radar image is a spatial distribution of the target's reflectivity mapped onto a range and Doppler plane. A range-Doppler image can be converted to a range and cross-range image if we have accurate knowledge of the scaling factor, which is determined by the rotation rate and the wavelength of the transmitted signal [11, 12].

An important factor of the image quality is its resolution (i.e., the ability to separate closely spaced scatterers in range and in cross range). The resolution along the radar LOS to the target is called the range (or down-range) resolution. The resolution transverse to the radar LOS is called the cross-range resolution. The minimum distance in the range Δr_r, and in the cross range Δr_{cr}, by which two point-scatterers can be separated, is the resolution of the image. A rectangle with sides Δr_r and Δr_{cr} is called a resolution cell. Range resolution is determined by the frequency bandwidth of the transmitter and the receiver. For an X-band radar operating at 10,000 MHz frequency, a bandwidth of 5% of the radar operation frequency (i.e., 500 MHz) can yield 1-ft range resolution. To obtain high cross-range

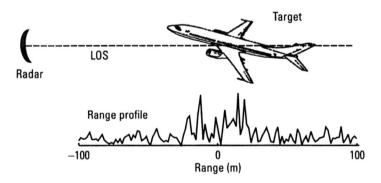

Figure 1.3 Radar range profile of an aircraft.

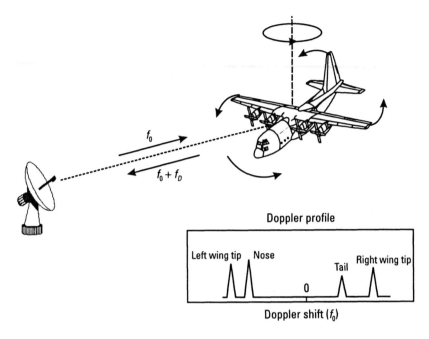

Figure 1.4 Doppler profile of an aircraft.

resolution a large antenna aperture is required. Usually, a synthetic aperture is utilized to synthesize a large antenna aperture. Synthetic aperture processing coherently combines signals obtained from sequences of small apertures at different aspect angles of a target to emulate the result that would be obtained from a large antenna aperture [10, 11].

Coherent processing maintains the relative phases of successive pulses. Thus, the phase from pulse to pulse is preserved and a phase correction can be applied to the returned signals to make them coherent for successive interpulse periods. If radar returns are processed coherently, the processed data retains both the amplitude and the phase information about the target. The amplitude is related to the radar cross section (a measure of the ability to reflect electromagnetic waves) of the target and the phase is related to the radial velocity of the target.

Imaging of moving targets using radar has been a major challenge. Techniques for high-resolution radar imaging are based on synthetic aperture processing described above. As long as there is a relative motion between the radar and the target, a synthetic aperture can be formed. A target can be considered as a set of individual point-scatterers, each with a radial velocity or Doppler frequency shift to the radar. Thus, the distribution of the radar reflectivity of the target can be measured by the Doppler frequency spectrum at each range gate, called the range bin or range cell, by taking the Fourier transform over the coherent processing interval (CPI) or imaging integration time. To use the Fourier transform properly, it is assumed that the frequency contents of the analyzed signal must be time-invariant. With this assumption, a long observation time results in high Doppler resolution. However, when the target moves, Doppler frequency shifts are time varying and the assumption of time-invariant Doppler frequency shifts is no longer valid. Thus, the Doppler spectrum becomes smeared, degrading the cross-range resolution, and the radar image becomes blurred. There are many methods, some called autofocusing and others called motion-compensation, for solving the problem of Doppler smearing and image blurring [13–22]. Most methods are Fourier-based approaches that attempt to flatten the Doppler spectra of individual scatterers by using sophisticated preprocessing approaches. Others use modern spectral analysis to achieve sharper images with shorter data samples [23–26]. In Chapter 5, we will introduce an image formation based on time-frequency transforms which can resolve the image blurring problem without resorting to sophisticated preprocessing algorithms. Time-frequency transform is a useful tool for radar imaging and signal analysis. In Chapters 3 through 8, we will introduce time-frequency transforms for detecting weak signals buried in noise, for analyzing radar back-scattering, for forming an image of maneuvering targets, for motion compensation, for moving target detection, and for micro-Doppler analysis.

In this chapter, we introduce basic concepts on radar back-scattering in Section 1.1. Then, we describe radar signals, noise, and the SNR in Section 1.2. In Section 1.3 we discuss the radar ambiguity function and matched filter, which is considered as a basic mathematical tool for signal design, analysis, and processing. In Section 1.4, we briefly introduce synthetic aperture radar imaging and its resolutions.

1.1 Electromagnetic Back-Scattering from Targets

The physical mechanism by which a transmitted radar signal is converted into a reflected signal due to the electromagnetic scattering from a target is a fundamental issue in understanding radar operation. This issue impacts the design of the radar waveform and its associated signal and image processing algorithms. It also determines how much of the target features can be extracted by the radar system. Most operational radars operate in the monostatic mode (i.e., the transmitter and the receiver are located physically at the same site).

Consequently, even though the target scatters the incident radar energy in all directions, only the energy that is scattered back to the radar transceiver is of interest (Figure 1.1). This is usually referred to as the back-scattered energy.

As one might expect, the electromagnetic back-scattering mechanisms are governed by the equations of Maxwell. They can be quite complicated for complex targets. A very simple model, however, called the point-scatterer model, has been used successfully in the radar community to approximately describe the behavior of back-scattered radar signal [27–30]. In this model, it is assumed that the scattering from a complex target can be approximately modeled as if it is emanating from a collection of localized scattering centers on the target. Therefore, the resulting relationship between the transmitted signal $s_T(t)$ and the received signal $s_R(t)$ can be written simply as

$$s_R(t) = \sum_n A_n s_T\left(t - \frac{2R_n}{c}\right) \tag{1.4}$$

where A_n represents the strength of the nth scattering center and R_n represents its location along the radar LOS or the so-called down-range direction. For a target that is stationary with respect to the radar, no Doppler frequency shift is involved in (1.4). Clearly, if the transmitted signal is a narrow pulse, then based on the point-scatterer model the reflected signal is comprised of a collection of pulses where the pulse locations indicate the spatial positions of the scattering centers on the target along the down-range direction. The strengths of the pulses are proportional to the strengths of the scattering centers. Thus, the received signal becomes a one-dimensional (1D) mapping of the prominent scattering centers on the target in the down-range direction. This is known as the range profile of the target and is an important feature space in radar signature diagnostic and target recognition applications [5]. For instance, the total extent of the range profile provides information on the length of the target along the radar LOS. The strong peaks in the range profile give the specific range locations of the strong scattering centers.

While the point-scatterer model is consistent with phenomenological observations, it can also be established more rigorously from first-principle electromagnetic theory. This is accomplished through the high-frequency approximation to Maxwell's equations, or ray optics. Through the early works of Luneburg and Kline [31] and Keller [32], and later Kouyoumjian and Pathak [33] and Lee and Deschamps [34], it was shown that the electromagnetic scattering from a complex target could be described by a set of highly localized ray phenomena. Each ray mechanism is attributable to a reflection or diffraction point on the target. For example, a single specular reflection point is used to describe the scattering from a smooth, curved surface. A diffraction point is used to describe the scattering from a sharp edge or a corner. Furthermore, the associated scattering amplitudes for a number of canonical configurations have been derived in closed form by electromagnetics researchers. Today, this knowledge base is generally called the geometrical theory of diffraction (GTD), which is a term first coined by Keller [32]. For radar applications, GTD provides the theoretical framework for the point-scatterer model. That is, the reflection and diffraction points on a target and their associated scattering amplitudes give rise to the point-scatterers observed in the actual range profiles and radar imagery.

In addition to providing a basis for the point-scatterer model, GTD is an important theory that allows us to examine the limits of the point-scatterer model. Since GTD was originally derived for electromagnetic fields in the frequency domain, we rewrite the point-scatterer model in (1.4) in the angular frequency domain $\omega = 2\pi f$ as

$$S_R(\omega) = S_T(\omega)\sum_n A_n \exp\left\{-j\omega \frac{2R_n}{c}\right\} \tag{1.5}$$

where $S_T(\omega)$ and $S_R(\omega)$ are Fourier transforms of $s_T(t)$ and $s_R(t)$, respectively. Note that in the above expression, each scattering mechanism has a constant amplitude and linear phase variation as a function of frequency. As a result, the incident pulse shape is fully preserved as it is scattered by each scattering center. Any scattering mechanism that satisfies the constant amplitude and linear phase condition above is usually called a "nondispersive" mechanism. Real scattering mechanisms, however, do in fact deviate from the idealized point-scatterer model. For instance, the scattering amplitude A_n derived from GTD for canonical conducting structures is, in general, weakly frequency dependent [35]. It has been shown that the frequency dependence of A_n is in the form of ω^{γ_n} where γ_n takes on half-integer values depending on the scatterer shape. Similarly, the phase of the individual exponential terms in (1.5) may exhibit nonlinear behavior as a function of frequency. This can occur in scatterers containing nonperfectly conducting materials or guided structures, such as inlet ducts, where the propagation of the electromagnetic energy differs from that in free space. As a result of such frequency dependencies that deviate from the idealized point-scatterer model, the return pulses in range are no longer identical in shape to the incident pulse. In general, they become much more spread out in range after the scattering process. When this occurs, we term the scattering process a dispersive one. The interpretation of dispersive scattering mechanisms in range is more difficult. In Chapter 4, we shall examine how dispersive features can be better revealed by time-frequency transforms.

1.2 Radar Signal and Noise

1.2.1 Signal Waveforms

In high range-resolution radar systems, in order to achieve high range resolution, signals having wide bandwidth are required. Widely used wideband signals include linear frequency modulated (LFM) signals and stepped frequency (SF) signals.

The LFM signal linearly changes its instantaneous carrier frequency within a single pulse as shown in Figure 1.5(a). The LFM signal with a Gaussian envelope can be expressed as

$$s_{LFM}(t) = (\alpha/\pi)^{1/4} \exp\{-\alpha t^2/2\} \exp\{j2\pi[f_0 + (\eta/2)t]t\} \quad (1.6)$$

where f_0 is the carrier frequency, η is the frequency-changing rate or chirp rate, and α determines the width of the Gaussian envelope.

The frequency spectrum of the LFM signal with a Gaussian envelope shown in Figure 1.5(b) can be derived as [36]

$$S_{LFM}(f) = \frac{(\alpha/\pi)^{1/4}}{(\alpha - j\eta)^{1/2}} \exp\left\{-\frac{2\pi^2\alpha(f-f_0)^2}{(\alpha^2 + \eta^2)} - j\frac{2\pi^2\eta(f-f_0)^2}{(\alpha^2 + \eta^2)}\right\}$$
$$(1.7)$$

Unlike the LFM signal, the SF signal achieves its wide bandwidth by sequentially changing the carrier frequency step-by-step over a number of pulses [Figure 1.6(a)]. Thus, the SF signal can be described by a sequence of pulses with increased carrier frequencies from one pulse to the next. The stepped carrier frequency can be expressed as

$$f_m = f_0 + (m - 1)\Delta f \quad (m = 1, 2, \ldots, M) \quad (1.8)$$

where Δf is the frequency step. The total bandwidth of the SF signal, $M\Delta f$, determines the radar range resolution. Because pulses are transmitted with a given pulse repetition interval (PRI) T_{PRI}, the SF signal can be expressed as

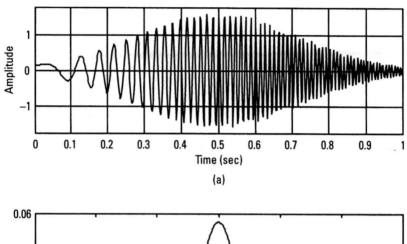

(a)

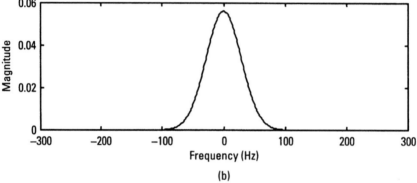

(b)

Figure 1.5 (a) An LFM waveform, and (b) its frequency spectrum.

$$s_{SF}(t) = \sum_{m=0}^{M-1} rect(t - mT_{PRI}, T) \exp\{-j2\pi(f_0 + m\Delta f)t\} \quad (1.9)$$

where the rectangular pulse is defined as

$$rect(t, T) = \begin{cases} 1 & |t| \le T/2 \\ 0 & |t| > T/2 \end{cases} \quad (1.10)$$

where $T \le T_{PRI}$ is the width of the rectangular pulse in time that determines the duty cycle of the pulse.

The Fourier transform of the rectangular pulse is $FT\{rect(t, T)\} = T\operatorname{sinc}(fT)$, where $\operatorname{sinc}(\cdot)$ is the sinc function defined by $\operatorname{sinc}(x) = \sin(\pi x)/(\pi x)$ (for $x \ne 0$) and $\operatorname{sinc}(x) = 1$ (for $x = 0$). Then, the Fourier transform of the time-shifted rectangular pulse $rect(t - \tau, T)$ becomes

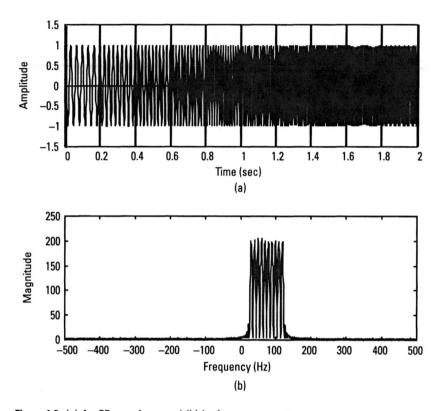

Figure 1.6 (a) An SF waveform, and (b) its frequency spectrum.

$$FT\{rect(t - \tau, T)\} = T\,\mathrm{sinc}(fT)\exp\{-j2\pi f\tau\} \qquad (1.11)$$

where τ is the time-shift. Therefore, the frequency spectrum of the SF signal can be derived by taking the Fourier transform of the signal $s_{SF}(t)$:

$$S_{SF}(f) = \sum_{m=0}^{M-1} T\,\mathrm{sinc}[(f-f_m)T]\exp\{-j2\pi(f-f_m)mT_{PRI}\}$$

$$\qquad (1.12)$$

where $f_m = f_0 + m\Delta f$ is the carrier frequency of the mth pulse. Figure 1.6(b) shows the frequency spectrum of the SF signal.

1.2.2 The SNR

Generally, SNR means the ratio of the intensity of the signal to the average intensity of the noise. The higher the SNR, the easier the signal detection.

There are two types of definitions of SNR: the average power SNR and the instantaneous power SNR.

The average power SNR is defined as the ratio of the average signal power to the average noise power. The average power is defined by

$$P_S = \frac{1}{T} \int_0^T s^2(t)dt \tag{1.13}$$

where T is the time duration of the signal $s(t)$. The average noise power is defined by

$$P_N = \int_{-\infty}^{\infty} (r_n - mean\{r_n\})^2 p(r_n)dr_n \tag{1.14}$$

where $r_n = \{n(t)\}$ is a random noise process and $p(r_n)$ is the probability density function of the random noise process r_n. Thus, the SNR becomes

$$SNR_{average} = \frac{P_S}{P_N} \tag{1.15}$$

For an additive white Gaussian noise (i.e., uniform power spectrum and Gaussian amplitude distribution) with zero-mean and variance $\sigma_{r_n}^2$, the average noise power is $P_N = \sigma_{r_n}^2$, and the average power SNR becomes

$$SNR_{average} = \frac{\frac{1}{T} \int_0^T s^2(t)dt}{\sigma_{r_n}^2} \tag{1.16}$$

The instantaneous power SNR is defined as the ratio of the instantaneous signal power to the average noise power:

$$SNR_{instant} = \frac{P_{instant}}{\sigma_{r_n}^2} \tag{1.17}$$

where instantaneous signal power is $P_{instant} = s^2(t)$. These two definitions of SNR are only suitable for linear systems because they do not take into account cross-terms between the signal and the noise. The matched filter introduced in Section 1.3.2 is based on the definition of the instantaneous power SNR to derive a linear system that outputs the maximum instantaneous peak power of the signal.

1.3 Radar Ambiguity Function and Matched Filter

1.3.1 Radar Ambiguity Function

Radar ambiguity function, first introduced by Woodward in 1953 [37], is a basic mathematical tool for signal design and analysis. It can be used for characterizing radar performance in target resolution and clutter rejection. The ambiguity function of a signal $s(t)$ is a 2D function in Doppler frequency shift f_D and time-delay τ and is defined as follows [38]:

$$\chi_S(\tau, f_D) = \int_{-\infty}^{\infty} s(t)s^*(t - \tau)\exp\{j2\pi f_D t\}dt \qquad (1.18)$$

$$= \int_{-\infty}^{\infty} S^*(f)S(f - f_D)\exp\{j2\pi f\tau\}df$$

where the asterisk refers to the conjugate, and $S(f)$ is the signal frequency spectrum. A high value of the ambiguity function indicates that it is difficult to resolve two nearby targets whose differences in the time delay and in the Doppler frequency shift are τ and f_D, respectively.

The ambiguity function can also be defined in a symmetrical form

$$\chi_s(\tau, f_D) = \int_{-\infty}^{\infty} s(t + \tau/2)s^*(t - \tau/2)\exp\{j2\pi f_D t\}dt \qquad (1.19)$$

$$= \int_{-\infty}^{\infty} S^*(f + f_D/2)S(f - f_D/2)\exp\{j2\pi f\tau\}df$$

From the above definition, we can easily see that the ambiguity function is a symmetric function:

$$\chi_s(\tau, f_D) = \chi_s^*(-\tau, -f_D) \qquad (1.20)$$

and its peak value is always at the center of the origin:

$$|\chi_s(\tau, f_D)| \leq |\chi_s(0, 0)| \qquad (1.21)$$

The peak of the ambiguity function $\chi_s(0, 0)$ means that it is impossible to resolve two nearby targets if their differences in the time delay and the Doppler frequency shift are all zeros. An ideal ambiguity function is a thumbtack-type function which has a peak value at $(\tau = 0, f_D = 0)$ and is near zero elsewhere. With the thumbtack-type ambiguity function, this means that two nearby targets can be perfectly resolved if their differences in the time delay and the Doppler frequency shift are not zeros. Of course, if $\tau = 0$ and $f_D = 0$, the ambiguity function has an infinite peak that makes two targets ambiguous.

Other properties of the ambiguity function include the following:

1. The ambiguity function of a scaled signal $s(\alpha t)$ is

$$s'(t) = s(\alpha t) \Rightarrow \chi_{s'}(\tau, f_D) = \frac{1}{|\alpha|}\chi_s\left(\alpha\tau, \frac{f_D}{\alpha}\right) \qquad (1.22)$$

2. The ambiguity function of a time-shifted signal $s(t - \Delta t)$ is

$$s'(t) = s(t - \Delta t) \Rightarrow \chi_{s'}(\tau, f_D) = \chi_s(\tau, f_D)\exp\{-j2\pi f_D \Delta t\} \qquad (1.23)$$

3. The ambiguity function of a frequency-modulated signal $s(t)\exp\{j2\pi ft\}$ is

$$s'(t) = s(t)\exp\{j2\pi ft\} \Rightarrow \chi_{s'}(\tau, f_D) = \chi_s(\tau, f_D)\exp\{-j2\pi f\tau\} \qquad (1.24)$$

If we set the Doppler shift to zero, the ambiguity function becomes the autocorrelation function of the signal $s(t)$

$$\chi(\tau, 0) = \int_{-\infty}^{\infty} s(t)s^*(t - \tau)dt \qquad (1.25)$$

For a rectangular pulse $rect(t, T)$ defined in (1.10), the ambiguity function as shown in Figure 1.7 is

$$\chi_{rect}(\tau, f_D) = \int_{-\infty}^{\infty} rect(t + \tau/2)rect^*(t - \tau/2)\exp\{j2\pi f_D t\}dt$$

$$= \int_{-(T-\tau)/2}^{(T-\tau)/2} \exp\{j2\pi f_D t\}dt \qquad (1.26)$$

$$= \begin{cases} (T - |\tau|)\operatorname{sinc}[f_D(T - |\tau|)] & \text{for } |\tau| \leq T \\ 0 & \text{for } |\tau| > T \end{cases}$$

For a Gaussian pulse

$$g(t) = e^{-\alpha t^2} \qquad (1.27)$$

the ambiguity function is also a Gaussian type:

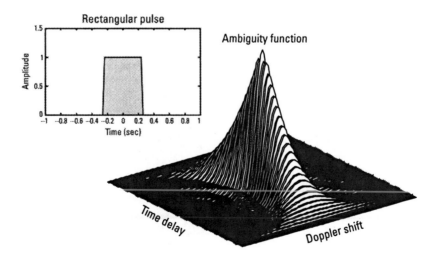

Figure 1.7 Ambiguity function of a rectangular pulse.

$$\chi_G(\tau, f_D) = \frac{1}{\sqrt{2}} \exp\{-\alpha(\tau^2 + f_D^2)/2\} \qquad (1.28)$$

For an LFM pulse

$$LFM(t) = \frac{1}{\sqrt{T}} rect\left(\frac{t - T/2}{T}\right) \exp\{j(2\pi f_0 t + \pi \eta t^2\} \qquad (1.29)$$

the ambiguity function as shown in Figure 1.8 becomes

$$\chi_s(\tau, f_D) = \begin{cases} (T - |\tau|) \, \text{sinc}[(f_D - \eta\tau)(T - |\tau|)] & |\tau| \leq T \\ 0 & otherwise \end{cases}$$

$$(1.30)$$

For an SF signal described in (1.9), its ambiguity function is shown in Figure 1.9, where the number of steps M = 10. The SF signal in the time domain and its time-frequency distribution are shown in Figures 1.9(a, b), respectively. Figures 1.9(c, d) are a surface plot and a contour plot of the ambiguity function of the SF signal, respectively. The details of the mathematical expression for the ambiguity function of the SF signal can be found in [39].

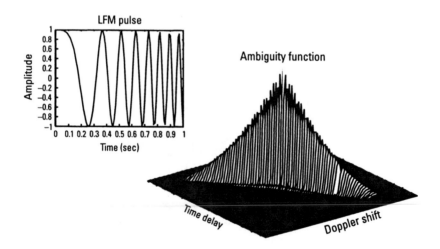

Figure 1.8 Ambiguity function of an LFM pulse.

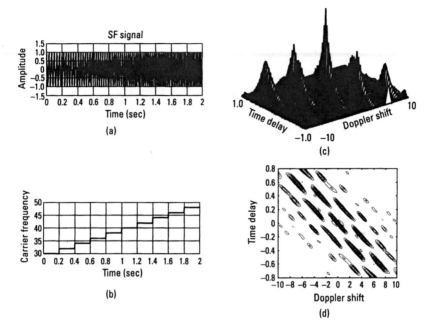

Figure 1.9 Ambiguity function of an SF signal: (a) SF signal; (b) time-frequency distribution; (c) surface of the ambiguity function; and (d) contour plot of the ambiguity function.

1.3.2 Matched Filter

A matched filter [40, 41] is a filter that provides the maximum output SNR when the signal is corrupted by white Gaussian noise. White noise means the power spectrum of the noise $P_n(f)$ is uniformly distributed over the entire frequency domain $(-\infty < f < \infty)$ (i.e., $P_n(f) = N_0/2$), and Gaussian noise indicates that the probability density function of the amplitude of the noise is a Gaussian distribution. Given a signal $s(t)$ in white Gaussian noise, the transfer function of the matched filter is the complex conjugate spectrum of the time-shifted signal $s(t + t_0)$ [40]:

$$H(f) = kS^*(f)\exp\{-j2\pi f t_0\} \tag{1.31}$$

and its impulse response is a mirror function of the input signal:

$$h(t) = \begin{cases} ks^*(t_0 - t), & (t \geq 0) \\ 0, & (t < 0) \end{cases} \tag{1.32}$$

where t_0 is a predetermined observation time and k is a scaling constant.

The matched filter has many important properties:

1. Among all linear filters, the matched filter is the only one that can provide maximum output SNR of $2E/N_0$, where $E = \int\limits_{-\infty}^{\infty} s^2(t)dt$ is the energy of the signal $s(t)$, and $N_0/2$ is the power spectrum density of the white noise.

2. An optimal filter matched to the signal $s(t)$ is also optimum to those signals that share the same waveform but different magnitude and time delay: $s'(t) = as(t - \tau)$ because of

$$
\begin{aligned}
H'(f) &= kS'^*(f)\exp\{-j2\pi ft'_0\} \\
&= akS^*(f)\exp\{-j2\pi f(t'_0 - \tau)\} \\
&= aH(f)\exp\{-j2\pi f[t'_0 - (t_0 + \tau)]\} = aH(f)
\end{aligned}
\tag{1.33}
$$

where $t'_0 = t_0 + \tau$. A matched filter to the signal $s(t)$ is, however, not optimum to those signals that have the same waveform but different frequency-shift $S'(f) = S(f + \nu)$. This is because the transfer function of the matched filter to the frequency-shift signal is

$$
H'(f) = kS^*(f + \nu)\exp\{-j2\pi ft_0\}
\tag{1.34}
$$

3. The matched filter is equivalent to a correlator. Because the impulse response of the matched filter is a mirror of the input signal, the output of the matched filter can be expressed as an autocorrelation function of the signal:

$$
s_{out}(t) = \int\limits_{-\infty}^{\infty} ks^*(t_0 - t)s(t - \tau)d\tau = kR_s(t - t_0)
\tag{1.35}
$$

4. The output of the matched filter to a signal is the Fourier transform of the power spectrum $|S(f)|^2$ of that signal:

$$s_{out}(t, t_0) = \int S(f)[kS^*(f) \exp\{-j2\pi f t_0\}] \exp\{j2\pi f t\} df$$

$$= \int kS(f)S^*(f) \exp\{j2\pi f(t - t_0)\} df \qquad (1.36)$$

$$= k \int |S(f)|^2 \exp\{j2\pi f(t - t_0)\} df$$

At an observation time t_0, which must be equal to or greater than the signal time duration, the output of the matched filter can achieve its maximum value.

For simplicity, the observation time is selected at the origin $t_0 = 0$. Then, the response of the matched filter $s_{out}(t, t_0)$ in (1.36) and the ambiguity function $\chi(\tau, f_D)$ in (1.18) are related by

$$s_{out}(t, 0) = \chi(t, 0) \qquad (1.37)$$

This means that the response of a matched filter can be derived from the ambiguity function by a cut along the line at $f_D = 0$.

1.3.3 Pulse Compression

For a high range-resolution radar with wideband waveforms, the matched filter actually functions as a pulse compression filter. For example, given an LFM signal with its frequency spectrum described in (1.7), according to (1.36), the output of the matched filter becomes

$$s_{out}(t, t_0) = \int k \frac{\left(\dfrac{\alpha}{\pi}\right)^{1/2}}{\alpha - j\beta} e^{-\frac{4\pi^2\alpha(f-f_0)^2}{(\alpha^2+\beta^2)}} e^{j2\pi f(t-t_0)} df$$

$$= k \frac{\left(\dfrac{\alpha}{\pi}\right)^{1/2}}{\alpha - j\beta} e^{j2\pi f_0(t-t_0)} \int e^{-\frac{4\pi^2\alpha}{(\alpha^2+\beta^2)}f^2} e^{j2\pi f(t-t_0)} df \quad (1.38)$$

$$= k \frac{\left(\dfrac{\alpha}{\pi}\right)^{1/2}}{\alpha - j\beta} \left(\frac{\alpha^2 + \beta^2}{8\pi^2\alpha}\right)^{1/2} e^{-\frac{\alpha^2+\beta^2}{16\pi^2\alpha}(t-t_0)^2} e^{j2\pi f_0(t-t_0)}$$

where we used the formula

$$\int e^{-Af^2} e^{j2\pi f(t-t_0)} df = \frac{1}{(2A)^{1/2}} e^{-\frac{1}{4A}(t-t_0)^2}$$

The matched filter output is plotted in Figure 1.10(b) and has a peak value at $t = t_0$. As described in (1.36), the output of the matched filter is the Fourier transform of the power spectrum of the signal. Therefore, for the purpose of pulse compression, we can either apply a matched filter to the signal or take the Fourier transform of the power spectrum of that signal as shown in Figure 1.10(c).

1.4 Synthetic Aperture Radar Imaging

Synthetic aperture radar (SAR), as an airborne or space-borne radar developed in the early 1950s, provides capabilities in generating radar image with

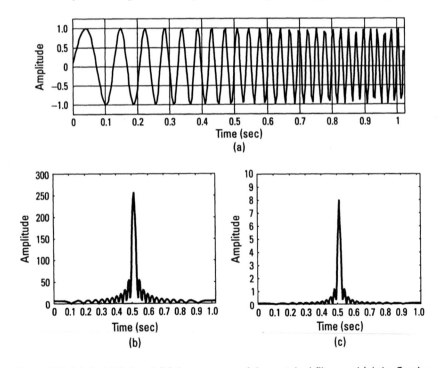

Figure 1.10 (a) An LFM signal; (b) the response of the matched filter; and (c) the Fourier transform of the power spectrum of the LFM signal.

fine features in addition to its position and velocity [10–12]. SAR images reconstructed from received signals are high-resolution maps of the spatial distribution of the reflectivity function of stationary surface targets and terrain. High-range resolution is obtained by using wide bandwidth of the transmitted waveform, and high cross-range resolution is achieved by coherently processing returned signals from a sequence of small apertures at different aspect angles to the radar to emulate a large aperture. If the radar is stationary and the target is moving, the angular motion of the target with respect to the radar can also be used to generate an image of the target. In this case, the radar is called an inverse synthetic aperture radar (ISAR) because it utilizes a geometrically inverse way (i.e., the radar is stationary and the target is moving) to image the target [9, 13–16]. In ISAR, cross-range resolution is determined by the Doppler resolution with a scaling factor. With a high Doppler resolution, differential Doppler shifts of adjacent scatterers on a target can be observed, and the distribution of the target's reflectivity can be obtained through the Doppler frequency spectrum. Conventional methods to obtain Doppler information are based on the Fourier transform, called the Fourier-based image formation.

1.4.1 Range Profile

With sufficient frequency bandwidth, it is possible to generate a 1D map of the target along the radar LOS, called the range profile or down-range profile as mentioned in Section 1.1. Similarly, by observing a target with relative motion with respect to the radar over a sufficient time interval, it is possible to generate a 1D cross-range map of the target.

A range profile is a range-compressed back-scattered signal. Since time delay is related to the distance from the radar to the target, the resulting radar signal, as a function of time, can be interpreted as a 1D mapping of the prominent scattering centers on the target along the radar LOS. In simple targets, a range profile typically consists of a number of distinct peaks that can be related spatially to the isolated scattering centers on the target.

1.4.2 Range Resolution

Radar range resolution defines the ability of resolving two point-targets within the same antenna beam, close together in the range domain. Because the delay-time τ of a radar signal returned from a target is related to the range R by $\tau = 2R/c$, the resolution in range is directly related to the resolution in delay-time. The range or down-range resolution Δr_r is determined by the bandwidth of the transmitted signal BW:

$$\Delta r_r = \frac{c}{2BW} \qquad (1.39)$$

1.4.3 Cross-Range Resolution

Doppler resolution refers to the ability of resolving two targets in the radial velocity. The Doppler resolution Δf_D is related to the coherent integration time T by

$$\Delta f_D = \frac{1}{T} \qquad (1.40)$$

The cross-range resolution Δr_{cr} is determined by the angle extent of the synthesized apertures during the coherent integration time: $\Delta r_{cr} = \frac{c}{2\Omega f_0 T}$, where f_0 is the frequency of the transmitted signal and Ω is the rotation rate of the target. A longer integration time may provide higher cross-range resolution, but causes phase-tracking errors, which can degrade the Doppler resolution and result in image blurring. Because the Doppler resolution Δf_D is inversely proportional to the image time T, the cross-range resolution is proportional to the Doppler resolution with a scaling factor:

$$\Delta r_{cr} = \left(\frac{c}{2\Omega f_0}\right)\Delta f_D \qquad (1.41)$$

where $2\Omega f_0 / c$ is called the scaling factor.

References

[1] Eaves, J. J., and E. K. Reedy (eds.), *Principles of Modern Radar*, New York: Van Nostrand Reinhold, 1987.

[2] Skolnik, M. I., *Introduction to Radar Systems*, Second Edition, New York: McGraw-Hill, 1980.

[3] Barton, D. K., *Modern Radar System Analysis*, Norwood, MA: Artech House, 1988.

[4] Hudson, S., and D. Psaltis, "Correlation Filters for Aircraft Identification from Radar Range Profiles," *IEEE Trans. Aerospace and Electronic Systems*, Vol. 29, No. 3, 1993.

[5] Li, H. J., and S. H. Yang, "Using Range Profiles as Feature Vectors to Identify Aerospace Objects," *IEEE Trans. Aerospace and Electronic Systems*, Vol. 41, No. 3, 1993, pp. 261–268.

[6]   Zyweck, A., and R. E. Bogner, "Radar Target Classification of Commercial Aircraft," *IEEE Trans. Aerospace and Electronic Systems*, Vol. 32, No. 2, 1996, pp. 598–606.

[7]   Chen, V. C., "Radar Range Profile Analysis with Natural Frame Time-Frequency Representation," *SPIE Proc. on Wavelet Applications*, Vol. 3078, 1997, pp. 433–448.

[8]   Knott, E. F., "Radar Cross Section." In *Aspects of Modern Radar*, E. Brookner (ed.), Norwood, MA: Artech House, 1988.

[9]   Rihaczek, A. W., and S. J. Hershkowitz, *Radar Resolution and Complex-Image Analysis*, Norwood, MA: Artech House, 1996.

[10]   Harger, R. O., *Synthetic Aperture Radar System*, New York: Academic Press, 1970.

[11]   Wehner, D. R., *High-Resolution Radar, Second Edition*, Norwood, MA: Artech House, 1994.

[12]   Carrara, W. G., R. S. Goodman, and R. M. Majewski, *Spotlight Synthetic Aperture Radar—Signal Processing Algorithms*, Norwood, MA: Artech House, 1995.

[13]   Ausherman, D. A., et al., "Developments in Radar Imaging," *IEEE Trans. Aerospace and Electronic Systems*, Vol. 20, No. 4, 1984, pp. 363–400.

[14]   Prickett, M. J., and C. C. Chen, "Principles of Inverse Synthetic Aperture Radar (ISAR) Imaging," *IEEE EASCON*, 1980, pp. 340–345.

[15]   Chen, C. C., and H. C. Andrews, "Target Motion Induced Radar Imaging," *IEEE Trans. Aerospace and Electronic Systems*, Vol. 16, No. 1, 1980, pp. 2–14.

[16]   Walker, J., "Range-Doppler Imaging of Rotating Objects," *IEEE Trans. Aerospace and Electronic Systems*, Vol. 16, No. 1, 1980, pp. 23–52.

[17]   Wahl, D. E., et al., "Phase Gradient Autofocus—A Robust Tool for High Resolution SAR Phase Correction," *IEEE Trans. Aerospace and Electronic Systems*, Vol. 30, No. 3, 1994, pp. 827–834.

[18]   Kirk, J. C., "Motion Compensation for Synthetic Aperture Radar," *IEEE Trans. Aerospace and Electronic Systems*, Vol. 11, 1975, pp. 338–348.

[19]   Steinberg, B. D., "Microwave Imaging of Aircraft," *Proc. IEEE*, Vol. 76, No. 12, 1988, pp. 1578–1592.

[20]   Wu, H., et al., "Translational Motion Compensation in ISAR Image Processing," *IEEE Trans. Image Processing*, Vol. 14, No. 11, 1995, pp. 1561–1571.

[21]   Baqai, A. B., and Y. Hua, "Matrix Pencil Methods for ISAR Image Reconstruction," *ICASSP*, 1993, pp. 473–476.

[22]   Werness, S. A., et al., "Moving Target Imaging Algorithm for SAR Data," *IEEE Trans. Aerospace and Electronic Systems*, Vol. 26, No. 1, 1990, pp. 57–67.

[23]   Odendaal, J. W., E. Barnard, and W. I. Pistorius, "Two-Dimensional Superresolution Radar Imaging Using the MUSIC Algorithm," *IEEE Trans. Antennas and Propagat.*, Vol. 42, No. 10, 1994, pp. 1386–1391.

[24]   Gupta, I. J., "High-Resolution Radar Imaging Using 2-D Linear Prediction," *IEEE Trans. Antenna and Propagat.*, Vol. 42, No. 1, 1994, pp. 31–37.

[25]   Li, J., and P. Stoica, "An Adaptive Filtering Approach to Spectral Estimation and SAR Imaging," *IEEE Trans. Signal Processing*, Vol. 44, No. 6, 1996, pp. 1469–1484.

[26] Wu, R., Z. S. Liu, and J. Li, "Time-Varying Complex Spectral Estimation with Applications to ISAR Imaging," *Proc. Asilomar Conference on Signals, Systems, and Computers*, Pacific Grove, CA, November 1998.

[27] Yu, W. P., L. G. To, and K. Oii, "N-Point Scatterer Model RCS/Glint Reconstruction from High-Resolution ISAR Target Imaging," *Proc. End Game Measurement and Modeling Conference*, Point Mugu, CA, January 1991, pp. 197–212.

[28] Hurst, M. P., and R. Mittra, "Scattering Center Analysis via Prony's Method," *IEEE Trans. Antennas Propagat.*, Vol. 35, August 1987, pp. 986–988.

[29] Carriere, R., and R. L. Moses, "High-Resolution Radar Target Modeling Using a Modified Prony Estimator," *IEEE Trans. Antennas Propagat.*, Vol. 40, January 1992, pp. 13–18.

[30] Bhalla, R., and H. Ling, "3-D Scattering Center Extraction Using the Shooting and Bouncing Ray Technique," *IEEE Trans. Antennas Propagat.*, Vol. AP-44, November 1996, pp. 1445–1453.

[31] Kline, M., and I. W. Kay, *Electromagnetic Theory and Geometrical Optics*, New York: Wiley, 1965.

[32] Keller, J. B., "Geometrical Theory of Diffraction," *J. Opt. Soc. Am.*, Vol. 52, January 1962, pp. 116–130.

[33] Kouyoumjian, R. G., and P. H. Pathak, "A Uniform Geometrical Theory of Diffraction for an Edge in a Perfectly Conducting Surface," *Proc. IEEE*, Vol. 62, November 1974, pp. 1448–1461.

[34] Lee, S. W., and G. A. Deschamps, "A Uniform Asymptotic Theory of EM Diffraction by a Curved Wedge," *IEEE Trans. Antennas Propagat.*, Vol. AP-24, January 1976, pp. 25–34.

[35] Potter, L. C., et al., "A GTD-Based Parametric Model for Radar Scattering," *IEEE Trans. Antennas Propagat.*, Vol. 43, October 1995, pp. 1058–1067.

[36] Cohen, L., *Time-Frequency Analysis*, Prentice Hall, 1995.

[37] Woodward, P. W., *Probability and Information Theory, with Applications to Radar*, Pergamon Press, 1953.

[38] Rihaczek, A. W., *Principles of High-Resolution Radar*, Mark Resources, Inc., 1977.

[39] Gill, G. S., and J. C. Huang, "The Ambiguity Function of the Step Frequency Radar Signal Processor," *Proc. CIE International Conference on Radar*, 1996, pp. 375–380.

[40] North, D. O., "Analysis of Factors which Determine Signal-to-Noise Discrimination in Pulsed Carrier Systems," *Proc. IEEE*, Vol. 51, 1963, pp. 1015–1028.

[41] Kelly, E. J., and R. P. Wishner, "Matched-Filter Theory for High-Velocity Targets," *IEEE Trans. Military Electr.*, Vol. 9, 1965, pp. 56–69.

2

Time-Frequency Transforms

Since its introduction in the early nineteenth century, the Fourier transform has become one of the most widely used signal-analysis tools across many disciplines of science and engineering. The basic idea of the Fourier transform is that any arbitrary signal (of time, for instance) can always be decomposed into a set of sinusoids of different frequencies. The Fourier transform is generated by the process of projecting the signal onto a set of basis functions, each of which is a sinusoid with a unique frequency. The resulting projection values form the Fourier transform (or the frequency spectrum) of the original signal. Its value at a particular frequency is a measure of the similarity of the signal to the sinusoidal basis at that frequency. Therefore, the frequency attributes of the signal can be revealed via the Fourier transform. In many engineering applications, this has proven to be extremely useful in the characterization, interpretation, and identification of signals.

While the Fourier transform is a very useful concept for stationary signals, many signals encountered in real-world situations have frequency contents that change over time. The most common example is music, where the harmonic content of the acoustic signal changes for different notes. In this case, it is not always best to use simple sinusoids as basis functions and characterize a signal by its frequency spectrum. Joint time-frequency transforms were developed for the purpose of characterizing the time-varying frequency content of a signal. The best-known time-frequency representation of a time signal dates back to Gabor [1] and is known as the short-time Fourier transform (STFT). It is basically a moving window Fourier transform. By examining the frequency content of the signal as the time window is moved, a 2D time-frequency distribution called the spectrogram is generated. The spectrogram contains information on the frequency content of the signal at different time instances. One well-known drawback of the STFT is the resolution limit imposed by the window function. A shorter time window results in better time resolution, but leads to worse frequency resolution, and vice versa. To overcome the resolution limit of the STFT, a wealth of alternative time-frequency representations have been proposed.

In this chapter, we provide an overview of various time-frequency transforms developed by researchers in the signal processing community. They are broadly divided into two classes: linear time-frequency transforms and quadratic (or bilinear) transforms. In Section 2.1, we first discuss linear time-frequency transforms. The discussion commences with the STFT and moves on to two other linear transforms, the continuous wavelet transform (CWT) and the adaptive time-frequency representation. In Section 2.2, we discuss quadratic time-frequency transforms. We begin with the Wigner-Ville distribution (WVD) and discuss Cohen's class and the time-frequency distribution series (TFDS). The main purpose of this chapter is to lay the groundwork for subsequent chapters on radar applications of time-frequency transforms. Emphasis is therefore placed on the application perspective. More detailed theoretical discussions on time-frequency transforms can be found in two excellent texts by Cohen [2] and Qian and Chen [3].

2.1 Linear Time-Frequency Transforms

We begin our discussion of linear time-frequency transforms with a review of the Fourier transform. The Fourier transform of a time signal $s(t)$ is defined as

$$S(\omega) = \int_{-\infty}^{\infty} s(t) \exp\{-j\omega t\} dt \qquad (2.1)$$

where $\omega = 2\pi f$ is the angular frequency. In the context of functional expansion, $S(\omega)$ can be interpreted as the projection of the signal onto a complex exponential function $\exp\{j\omega t\}$ at angular frequency ω. Since the set of exponentials forms an orthogonal basis set, the original function can be constructed from the projection values by the process of

$$s(t) = \frac{1}{2\pi} \int_{-\infty}^{\infty} S(\omega) \exp\{j\omega t\} d\omega \qquad (2.2)$$

which is the inverse Fourier transform of $S(\omega)$. A well-known property of the Fourier transform pair $s(t)$ and $S(\omega)$ is the uncertainty principle. It states that the time duration Δ_t of $s(t)$ and the frequency bandwidth Δ_ω of $S(\omega)$ are related by

$$\Delta_t \Delta_\omega \geq \frac{1}{2} \tag{2.3}$$

where

$$\Delta_t = \left[\frac{\displaystyle\int_{-\infty}^{\infty} (t - \mu_t)^2 |s(t)|^2 \, dt}{\displaystyle\int_{-\infty}^{\infty} |s(t)|^2 \, dt} \right]^{1/2}$$

$$\Delta_\omega = \left[\frac{\displaystyle\int_{-\infty}^{\infty} (\omega - \mu_\omega)^2 |S(\omega)|^2 \, d\omega}{\displaystyle\int_{-\infty}^{\infty} |S(\omega)|^2 \, d\omega} \right]^{1/2}$$

and the mean time μ_t and mean frequency μ_ω are defined as

$$\mu_t = \frac{\displaystyle\int_{-\infty}^{\infty} t \, |s(t)|^2 \, dt}{\displaystyle\int_{-\infty}^{\infty} |s(t)|^2 \, dt}$$

$$\mu_\omega = \frac{\displaystyle\int_{-\infty}^{\infty} \omega \, |S(\omega)|^2 \, d\omega}{\displaystyle\int_{-\infty}^{\infty} |S(\omega)|^2 \, d\omega}$$

Thus, the larger the time duration of $s(t)$, the smaller the frequency bandwidth of $S(\omega)$. Conversely, the larger the frequency bandwidth of $S(\omega)$, the shorter the time duration of $s(t)$.

When we use (2.1) to estimate the frequency spectrum of a signal, we assume that the frequency content of the signal is relatively stable during the observation time interval. If the frequency content changes with time, it is not possible to monitor clearly how this variation takes place as a function of time. The reason can be attributed to the nature of the complex sinusoidal basis, which is of infinite duration in time. While the frequency spectrum can still be used to uniquely represent the signal, it does not adequately reflect the actual characteristics of the signal. In the following three subsections, three linear time-frequency transforms (viz., STFT, the CWT, and the adaptive time-frequency representation) are presented. They can be considered as a generalization of the Fourier transform with alternative basis sets that can better reflect the time-varying nature of the signal frequency spectrum.

2.1.1 The STFT

The most standard approach to analyze a signal with time-varying frequency content is to split the time-domain signal into many segments, and then take the Fourier transform of each segment (see Figure 2.1). This is known as the STFT operation and is defined as

$$STFT(t, \omega) = \int s(t')w(t' - t) \exp\{-j\omega t'\}dt' \qquad (2.4)$$

This operation (2.4) differs from the Fourier transform only by the presence of a window function $w(t)$. As the name implies, the STFT is generated by taking the Fourier transform of smaller durations of the original data. Alternatively, we can interpret the STFT as the projection of the function $s(t')$ onto a set of bases $w^*(t' - t) \exp\{j\omega t'\}$ with parameters t and ω. Since the bases are no longer of infinite extent in time, it is possible to monitor how the signal frequency spectrum varies as a function of time. This is accomplished by the translation of the window as a function of time t, resulting in a 2D joint time-frequency representation $STFT(t, \omega)$ of the original time signal. The magnitude display $|STFT(t, \omega)|$ is called the spectrogram of the signal. It shows how the frequency spectrum (i.e., one vertical column of the spectrogram) varies as a function of the horizontal time axis.

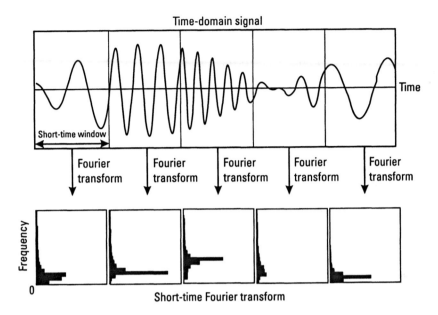

Figure 2.1 Illustration of the STFT.

The definition of the STFT can also be expressed in the frequency domain by manipulating (2.4), with the result

$$STFT(t,\,\omega) = \frac{1}{2\pi}\exp\{-j\omega t\}\int S(\omega')W(\omega - \omega')\exp\{j\omega't\}d\omega' \quad (2.5)$$

Here $W(\omega)$ is the Fourier transform of $w(t)$. The dual relationship between (2.4) and (2.5)[1] is apparent (i.e., the time-frequency representation can be generated via a moving window in time or a moving window in frequency). In addition, we make the following observations: (1) Signal components with durations shorter than the duration of the window will tend to get smeared out [i.e., the resolution in the time domain is limited by the width of the window $w(t)$]. Similarly, the resolution in the frequency domain is limited by the width of the frequency window $W(\omega)$. (2) The window width in time and the window width in frequency are inversely proportional to each other by the uncertainty principle. Therefore, good resolution in time (small time window) necessarily implies poor resolution in frequency (large frequency window). Conversely, good resolution in fre-

---

1. Equation (2.5) has also been referred to as the running-window Fourier transform [4].

quency implies poor resolution in time. (3) The window width in each domain remains fixed as it is translated. This results in a fixed resolution across the entire time-frequency plane. Figure 2.2 shows the basis functions of the STFT and the resulting fixed-resolution cells in the time-frequency plane.

So far, we have not discussed the specific shape of the window function. In general, to cut down on sidelobe interference in the spectrogram, the window function should taper to zero smoothly. Examples of window functions include Hamming, Hanning, Kaiser-Bessel, and Gaussian windows. An STFT using a Gaussian window function is sometimes called the Gabor transform [1]. If we let

$$w(t) = \frac{1}{\pi^{1/4}\sqrt{\sigma}} \exp\left\{-\frac{t^2}{2\sigma^2}\right\} \tag{2.6}$$

the corresponding frequency window is

$$W(\omega) = (2\sigma)^{1/2}\pi^{1/4} \exp\left\{-\frac{\sigma^2\omega^2}{2}\right\} \tag{2.7}$$

From (2.3), we have $\mu_t = 0$, $\mu_\omega = 0$, $\Delta_t = \sigma/\sqrt{2}$, $\Delta_\omega = 1/(\sqrt{2}\sigma)$, and $\Delta_t\Delta_\omega = 1/2$. We can see that the uncertainty equality in (2.3) holds for the

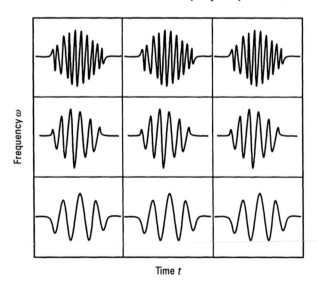

Figure 2.2 Basis functions and the resulting fixed-resolution cells of the STFT.

Gaussian function. Therefore, the Gaussian window function achieves the best time-frequency product among all the possible window functions.

Figure 2.3 shows an example of a signal containing four nonoverlapping, finite-duration sinusoids. Figure 2.3(a) is the time-domain waveform and Figure 2.3(b) shows its corresponding frequency spectrum. Although the four frequencies are well resolved, their time duration information cannot be seen in the frequency domain. Figure 2.3(c) is the STFT spectrogram generated using a Hanning window of 32 points. It shows both the frequency locations and time durations of the four signal components. Figure 2.3(d) is the spectrogram obtained by using a longer time window of 128 points. As expected, a longer time window results in better frequency localization in the time-frequency plane, at the expense of worse time resolution. These results (as well as subsequent examples in this chapter) were generated using the demonstration version of the Joint Time-Frequency Analyzer developed by the National Instruments Corporation [3].

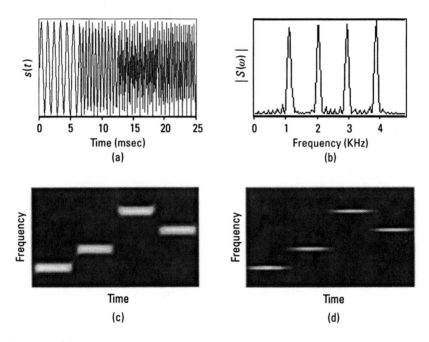

Figure 2.3 (a) A test signal in time consisting of four nonoverlapping, finite-duration sinusoids; (b) its frequency spectrum obtained via the Fourier transform; (c) its spectrogram obtained using STFT with a Hanning window of width 32 points; and (d) its spectrogram with a window of width 128 points. (Plots obtained using the demonstration version of the Joint Time-Frequency Analyzer developed by the National Instruments Corporation [3].)

2.1.2   The CWT

As described in the last section, the spectrogram generated by the STFT is limited in resolution by the extent of the sliding window function. A smaller time window results in better time resolution, but leads to worse frequency resolution, and vice versa. Contrary to the fixed resolution of the STFT, the wavelet transform is a time-frequency representation capable of achieving variable resolution in one domain (either time or frequency) and multiresolution in the other domain [5–8]. The CWT of a signal $s(t)$ can be defined as

$$CWT(t, \omega) = \left(\frac{\omega}{\omega_0}\right)^{1/2} \int s(t')\psi^* \left(\frac{\omega}{\omega_0}(t' - t)\right) dt' \qquad (2.8)$$

$\psi(\cdot)$ is usually termed the "mother wavelet" in wavelet theory. The ratio (ω_0/ω) is the scale parameter and the resulting 2D magnitude display of the above expression is called the scalogram. Let us assume that the mother wavelet is centered at time zero and oscillates at frequency ω_0. Essentially, (2.8) can be interpreted as a decomposition of the signal $s(t')$ into a family of shifted and dilated wavelets $\psi[(\omega/\omega_0)(t' - t)]$. The wavelet basis function $\psi[(\omega/\omega_0)(t' - t)]$ has variable width according to ω at each time t. The $\psi[(\omega/\omega_0)(t' - t)]$ is wide for small ω and narrow for large ω. By shifting $\psi(t')$ at a fixed parameter ω, the (ω_0/ω)-scale mechanisms in the time response $s(t')$ can be extracted and localized. Alternatively, by dilating $\psi(t')$ at a fixed t, all of the multiscale events of $s(t')$ at t can be analyzed according to the scale parameter (ω_0/ω). This is the multiresolution property of the wavelet transform and is an advantage over the STFT for analyzing multiscale signals.

The wavelet transform can also be carried out on the inverse Fourier transform $S(\omega)$ of the signal $s(t)$

$$CWT(t, \omega) = \frac{(\omega_0/\omega)^{1/2}}{2\pi} \int S(\omega')\Psi^* \left(\frac{\omega_0}{\omega}\omega'\right) \exp\{j\omega't\} d\omega' \qquad (2.9)$$

where $\Psi(\omega')$ is the Fourier transform of $\psi(t')$. Notice that (2.9) is essentially the Fourier transform of $S(\omega')\Psi^*[(\omega_0/\omega)\omega']$. By comparing (2.9) and (2.5), we observe that $\Psi^*(\omega')$ is similar to the frequency window function $W(\omega')$ in the running window Fourier transform. However, $\Psi(\omega')$ must satisfy the "admissibility condition" in wavelet theory, namely, $\Psi(0) = 0$, (i.e., it contains no dc components). To satisfy this condition, $\Psi(\omega')$ can be thought of as a shifted window function with a center frequency of ω_0. By changing ω, $\Psi[(\omega_0/\omega)\omega']$ is shifted to ω' and the width of the window is dilated by the factor (ω/ω_0). The ratio between the window width and the window center (or the Q-factor of the window function) remains fixed for all ω values. This is the constant-Q property of the wavelet filter and is in contrast to the STFT where the window width does not change as it is being shifted.

Figure 2.4 illustrates the basis functions in the CWT and the resulting time-frequency grid. Note that both the CWT and the STFT can be interpreted as the decomposition of the time signal $s(t)$ into a family of basis functions that determine the properties of the transform. The STFT and the CWT are similar to each other in that they both use finite basis functions. This is in contrast to the Fourier transform, which uses bases of infinite extent. As is shown in Figure 2.4, however, the width of the basis function in the CWT changes according to the frequency parameter, leading to variable resolution of the time-frequency plane.

2.1.3 Adaptive Time-Frequency Representation

Wavelet use is a step toward variable resolution in the time-frequency plane. However, it is still rather rigid in its particular form of the time-frequency

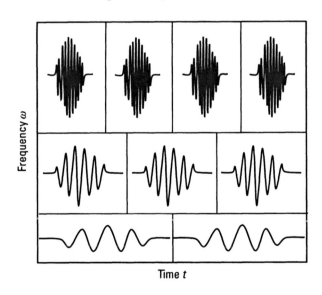

Figure 2.4 Basis functions and the resulting variable-resolution cells of the CWT.

grid. More flexible resolution in the time-frequency plane to accommodate components of the signal with different resolutions is sometimes desirable. Several signal-adaptive time-frequency representations have been proposed in the literature for this purpose, the best known of which are the adaptive Gaussian representation [9] and the matching pursuit algorithm [10]. The adaptive spectrogram (ADS), which we will discuss here, uses adaptive normalized Gaussian functions to represent the signal. In the algorithm, the time and frequency resolutions, as well as the time-frequency centers, are adjusted to best match the signal. The objective of this method is to expand a signal $s(t)$ in terms of normalized Gaussian functions $h_p(t)$ with an adjustable standard deviation σ_p and a time-frequency center (t_p, ω_p) as follows:

$$s(t) = \sum_{p=1}^{\infty} B_p h_p(t) \qquad (2.10)$$

where

$$h_p(t) = (\pi\sigma_p^2)^{-1/4} \exp\left\{-\frac{(t - t_p)^2}{2\sigma_p^2}\right\} \exp\{j\omega_p t\} \qquad (2.11)$$

Note that the modulated Gaussian basis has a dual form in its Fourier transform representation

$$H_p(\omega) = (\pi(1/2\pi\sigma_p)^2)^{-1/4} \exp\left\{-\frac{(\omega - \omega_p)^2}{2(1/\sigma_p)^2}\right\} \exp\{-j(\omega - \omega_p)t_p\}$$

$$(2.12)$$

Therefore, these basis functions have a time-frequency extent given by σ_p and $(1/\sigma_p)$, respectively (see Figure 2.5).

The coefficients B_p are found one at a time by an iterative procedure. We begin at the stage $p = 1$ and choose the parameters σ_p, t_p, and ω_p such that $h_p(t)$ is the basis with the maximum projection onto the signal

$$B_p = \max_{\sigma_p, t_p, \omega_p} \int s_{p-1}(t) h_p^*(t) dt \qquad (2.13)$$

where $s_0(t) = s(t)$. For $p > 1$, $s_p(t)$ is the remainder after the orthogonal projection of $s_{p-1}(t)$ onto $h_p(t)$ has been removed from the signal

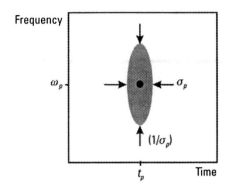

Figure 2.5 Time-frequency representation of a modulated Gaussian basis function centered at (t_p, ω_p) with standard deviation σ_p. (*Source:* [11] © 1997 IEEE.)

$$s_p(t) = s_{p-1}(t) - B_p(t)h_p(t) \qquad (2.14)$$

This procedure is iterated to generate as many coefficients as needed to accurately represent the original signal.

Several comments can be made about the adaptive Gaussian representation. First, it can be shown that the norm of the residue monotonically decreases and converges to zero. Therefore adding a new term in the series does not affect the previously selected parameters. Second, because this representation is adaptive, it will generally be concentrated in a very small subspace. As a result, we can use a finite summation of the terms in (2.10) to approximate the signal with a small residual error. Also, since random noise is in general distributed uniformly in the entire time-frequency space, this subspace representation actually increases the signal-to-noise ratio. (Chapter 3 discusses the denoising issue in detail.) Finally, the major difficulty in implementing this algorithm is the determination of the optimal elementary function at each stage. One implementation strategy is to start with a large σ_p and scan the data in frequency and time for a peak. We then divide σ_p by two and find the new peak. This procedure is continued until the standard deviation is small enough (as shown in Figure 2.6). We then select the highest peak and extract the residual using (2.14). It should be pointed out that the fast Fourier transform can be used during this search procedure to obtain the coefficients for all the frequency centers at once, speeding up a search that would otherwise be very time consuming.

The result of applying the adaptive Gaussian extraction can be effectively displayed in the time-frequency plane using the so-called ADS:

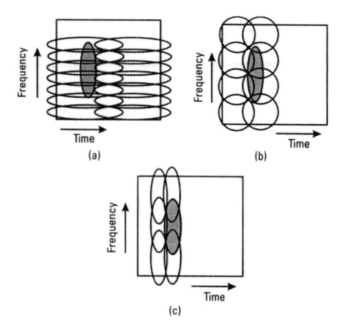

Figure 2.6 Illustration of the search strategy for the adaptive Gaussian representation: (1) start with large σ basis and locate the time-frequency position of the peak; (2) divide σ and the search region by two and repeat the search; and (3) repeat this procedure until the highest peak is found. (*Source:* [11] © 1997 IEEE.)

$$ADS(t,\ \omega) = 2\sum_p |B_p|^2 \exp\left[-\frac{(t - t_p)^2}{\sigma_p^2} - \sigma_p^2(\omega - \omega_p)^2 \right] \quad (2.15)$$

This representation is obtained by calculating the WVD (to be discussed in Section 2.2.1) of (2.10) and then deleting the cross terms. It can be shown that the energy contained in the ADS is identical to the energy contained in the signal. Therefore it can be considered as a signal energy distribution in the time-frequency domain. It is also nonnegative, free of cross-term interference, and of high resolution. Figure 2.7 shows the ADS of the test signal shown in Figure 2.3(a).

Further extension of the Gaussian basis functions to include other higher-order phase terms such as chirps have also been reported in [12, 13].

2.2 Bilinear Time-Frequency Transforms

The power spectrum of a signal $s(t)$ is the magnitude square of its Fourier transform, $|S(\omega)|^2$. It can also be expressed as the Fourier transform of the autocorrelation function of $s(t)$

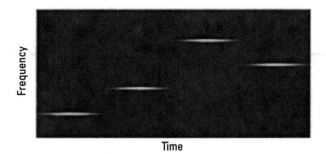

Figure 2.7 The ADS of the test signal with four nonoverlapping, finite-duration sinusoids.

$$|S(\omega)|^2 = \int R(t')e^{-j\omega t'}dt' \qquad (2.16)$$

where the autocorrelation function is given by

$$R(t') = \int s(t)s^*(t - t')dt \qquad (2.17)$$

The power spectrum indicates how the signal energy is distributed in the frequency domain. While the Fourier transform $S(\omega)$ is a linear function of $s(t)$, the power spectrum is a quadratic function of $s(t)$. Therefore, time-frequency distributions derived directly from the Fourier transform, such as those discussed in Section 2.1, can be classified as linear transforms, while it is customary to call those distributions derived from the power spectrum quadratic (or bilinear) time-frequency distributions. The main impetus for quadratic time-frequency distribution is to define an appropriate time-dependent power spectrum. In this section, we shall discuss three such time-frequency transforms, the WVD, Cohen's class, and the TFDS.

2.2.1 The WVD

The most basic of the quadratic time-frequency representations, the WVD, was first developed in quantum mechanics by Wigner in 1932 [14] and later introduced for signal analysis by Ville [15]. In the WVD, a time-dependent autocorrelation function is chosen as

$$R(t, t') = s\left(t + \frac{t'}{2}\right)s^*\left(t - \frac{t'}{2}\right) \qquad (2.18)$$

The WVD of $s(t)$ is then defined as the Fourier transform of this time-dependent autocorrelation function

$$WVD(t, \omega) = \int s\left(t + \frac{t'}{2}\right)s^*\left(t - \frac{t'}{2}\right)\exp\{-j\omega t'\}dt' \qquad (2.19)$$

The WVD can also be defined based on the Fourier transform of $s(t)$ as follows:

$$WVD(t, \omega) = \frac{1}{2\pi}\int S\left(\omega + \frac{\omega'}{2}\right)S^*\left(\omega - \frac{\omega'}{2}\right)\exp\{j\omega' t\}d\omega'$$

$$(2.20)$$

The WVD has a number of desirable properties that make it a good indicator of how the energy of the signal can be viewed as a function of time and frequency. First the WVD of any signal is always real. Second, it satisfies the *time marginal condition*

$$\frac{1}{2\pi}\int WVD(t, \omega)d\omega = |s(t)|^2 \qquad (2.21)$$

That is, by summing the time-frequency distribution over all frequencies, we obtain the instantaneous energy of the signal at a particular time instance. Similarly, the WVD also satisfies the *frequency marginal condition* given by

$$\int WVD(t, \omega)dt = |S(\omega)|^2 \qquad (2.22)$$

In this case, by summing the time-frequency distribution over all time, we obtain the power spectrum of the signal at a particular frequency. Third, the WVD satisfies the *instantaneous frequency property*. If $s(t) = A(t)\exp\{j\alpha(t)\}$, then the average frequency for a given time t is

$$\mu_{\omega/t} = \frac{\displaystyle\int \omega\, WVD(t, \omega)d\omega}{\displaystyle\int WVD(t, \omega)d\omega} = \frac{d}{dt}\alpha(t) \qquad (2.23)$$

That is, the mean frequency computed from the WVD is equal to the derivative of the phase (i.e., the mean instantaneous frequency of the signal). Similarly, the WVD also satisfies the *group delay property*. If $S(\omega) = B(\omega) \exp\{j\beta(\omega)\}$, then the group delay for a given ω is

$$\mu_{t/\omega} = \frac{\int t WVD(t, \omega)dt}{\int WVD(t, \omega)dt} = -2\pi\frac{d}{d\omega}\beta(\omega) \qquad (2.24)$$

It implies that the mean time computed from the WVD is equal to the derivative of the spectral phase (i.e., the group delay of the signal).

Although the WVD has many nice properties and gives nearly the best resolution among all the time-frequency techniques, its main drawback comes from cross-term interference. Simply put, the WVD of the sum of two signals is not the sum of their WVDs. If $s = s_1 + s_2$, it can be shown that

$$WVD_s(t, \omega) = WVD_{s_1}(t, \omega) + WVD_{s_2}(t, \omega) + 2\mathrm{Re}\{WVD_{s_1 s_2}(t, \omega)\}$$
$$(2.25)$$

where the last term is the cross WVD of s_1 and s_2 given by

$$WVD_{s_1 s_2}(t, \omega) = \int s_1\left(t + \frac{t'}{2}\right)s_2^*\left(t - \frac{t'}{2}\right)\exp\{-j\omega t'\}dt' \qquad (2.26)$$

As a result, if a signal contains more than one component in the joint time-frequency plane, its WVD will contain cross terms that occur halfway between each pair of autoterms. The magnitude of these oscillatory cross terms can be twice as large as the autoterms and yet they do not possess any physical meaning. Figure 2.8 shows an example of a signal containing four finite-duration sinusoids shown earlier in Figure 2.3(a). We can see that even though the WVD has very good time-frequency localization, there are cross-term interference terms between every pair of signal components. This drawback severely hinders the usefulness of the WVD for detecting signal characteristics in the time-frequency plane.

2.2.2 Cohen's Class

In addition to the WVD, a number of bilinear distributions have also been proposed by researchers for time-frequency signal analysis [16–18]. In 1966,

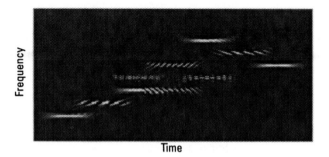

Figure 2.8 The WVD of the test signal with four nonoverlapping, finite-duration sinusoids.

Cohen showed that all these existing time-frequency distributions could be written in a generalized form [19]. Moreover, this general form can be used to facilitate the design of new time-frequency transforms. This class of transforms is now known simply as Cohen's class. We shall describe the general form of Cohen's class, followed by two well-known members of the class for reducing the cross-term interference problem in the WVD.

The general form of Cohen's class is defined as

$$C(t, \omega) = \int\int s\left(u + \frac{t'}{2}\right)s^*\left(u - \frac{t'}{2}\right)\phi(t - u, t')\exp\{-j\omega t'\}dudt'$$

(2.27)

The Fourier transform of $\phi(t, t')$, denoted as $\Phi(\theta, t')$, is called the kernel function. It can easily be seen that if $\Phi(\theta, t') = 1$, then $\phi(t, t') = \delta(t)$ and (2.27) reduces to the WVD defined in (2.19). Therefore, the WVD is a member of Cohen's class. More generally, other types of kernel functions can be designed to reduce the cross-term interference problem of the WVD. Two such time-frequency distributions are the Choi-Williams distribution (CWD) and the cone-shaped distribution (CSD).

The CWD [20] uses as its kernel function

$$\Phi(\theta, t') = \exp\{-\alpha(\theta t')^2\}$$

(2.28)

Along the θ-axis and the t'-axis, the kernel function is identically one while away from the two axes, the function decays with the damping controlled by α. The inverse Fourier transform of $\Phi(\theta, t')$ is given by

$$\phi(t, t') = \frac{1}{\sqrt{4\pi\alpha(t')^2}} \exp\left\{-\frac{t^2}{4\alpha(t')^2}\right\} \qquad (2.29)$$

and the CWD is defined as

$$CWD(t, \omega) = \int\int \frac{1}{\sqrt{4\pi\alpha(t')^2}} \qquad (2.30)$$

$$\exp\left\{-\frac{(t-u)^2}{4\alpha(t')^2}\right\} s\left(u + \frac{t'}{2}\right) s^*\left(u - \frac{t'}{2}\right) \exp\{-j\omega t'\}\,du\,dt'$$

Note that the kernel function is essentially a low-pass filter in the θ-t' plane. It preserves all cross terms that are on the θ-axis and t'-axis. As a result, the CWD usually contains strong horizontal and vertical cross terms in the time-frequency plane. Figure 2.9 shows the CWD of the same test signal containing four finite-duration sinusoids. It preserves the property of the WVD while reducing cross-term interference.

The CSD was introduced by Zhao, Atlas, and Marks [21]. Its name comes from the definition of a cone-shaped $\phi(t, t')$

$$\phi(t, t') = \begin{cases} g(t'), & |t'| \ge 2|t| \\ 0 & otherwise \end{cases} \qquad (2.31)$$

which is confined to the region bounded by lines $t' = 2t$ and $t' = -2t$. In this case, the corresponding kernel function is of the form

$$\Phi(\theta, t') = g(t')|t'|\operatorname{sinc}\left(\frac{\theta t'}{2}\right) \qquad (2.32)$$

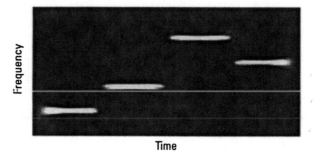

Figure 2.9 The CWD of the test signal with four nonoverlapping, finite-duration sinusoids.

For example, if we choose $g(t') = (1/|t'|) \exp\{-\alpha(t')^2\}$, then the kernel function is $\Phi(\theta, t') = \exp\{-\alpha(t')^2\} \operatorname{sinc}(\theta t'/2)$. In this case, the kernel function is one along the θ-axis and $\exp\{-\alpha(t')^2\}$ along the t'-axis where α controls the decay. Figure 2.10 shows the CSD of the same test signal. Again, the CSD reduces cross-term interference while nearly maintaining the resolution of the WVD.

2.2.3 The TFDS

Another approach to overcoming the cross-term interference problem of the WVD is the TFDS, proposed by Qian and Chen [22]. They suggested that if the WVD can be decomposed into a sum of localized and symmetric functions, it may be possible to suppress cross-term interference by selecting only the low-order harmonics. This is accomplished by first decomposing the original signal into the Gabor expansion

$$s(t) = \sum_m \sum_n C_{m,n} h_{m,n}(t) \tag{2.33}$$

where

$$h_{m,n}(t) = (\pi\sigma^2)^{-1/4} \exp\left\{\frac{(t - m\Delta t)^2}{2\sigma^2} + jn\Delta\omega t\right\} \tag{2.34}$$

are time-shifted and frequency-modulated Gaussian basis functions. In the above expression, m and Δt are respectively the time sampling index and time sampling interval, while n and $\Delta\omega$ are the sampling index and sampling interval in frequency. In other words, C_{mn} represents the STFT of the function $s(t)$ using a Gaussian window and evaluated on a sampled grid.

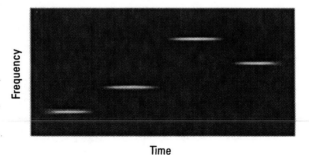

Figure 2.10 The CSD of the test signal with four nonoverlapping, finite-duration sinusoids.

By taking the WVD of both sides of (2.33), we obtain

$$WVD(t, \omega) = \sum_{mn} \sum_{m'n'} C_{m,n} C^*_{m',n'} WVD_{h,h'}(t, \omega) \qquad (2.35)$$

where $WVD_{h,h'}$ denotes the WVD between any pair of basis functions and is available in closed form. Next, the above expression can be regrouped based on the "interaction distance"

$$D = |m - m'| + |n - n'| \qquad (2.36)$$

between the pairs of bases (m, n) and (m', n'). This results in what is termed the TFDS, also called the Gabor spectrogram:

$$TFDS_D(t, \omega) = \sum_{mn} |C_{m,n}|^2 WVD_{h,h'}(t, \omega) \qquad (D = 0 \text{ terms})$$

$$+ \sum_{mn} \sum_{m'n'} C_{m,n} C^*_{m',n'} WVD_{h,h'}(t, \omega) \quad (D = 1 \text{ terms})$$

$$+ \sum_{mn} \sum_{m'n'} C_{m,n} C^*_{m',n'} WVD_{h,h'}(t, \omega) \quad (D = 2 \text{ terms})$$

$$+ \ldots \qquad (2.37)$$

Clearly, if we take all the terms in the series ($D = \infty$), the right-hand side of (2.37) converges to the WVD of the original signal. This yields the best resolution but is plagued by cross-term interference. At the other extreme, if we take only the self-interaction terms in the series ($D = 0$), the resulting right-hand side is equivalent to the spectrogram of the signal using a Gaussian window function. It has no cross-term interference problem but has the worst resolution. As the order D increases, we gain in resolution but pay a price in cross-term interference. It is often possible to balance the resolution against cross-term interference by adjusting the order D. The optimal value for D was reported to be around 2 to 4.

Figure 2.11 shows the effect of the order D on the frequency-hopping signal discussed in earlier examples. For $D = 0$ [Figure 2.11(a)] the signal has the least time-frequency resolution, but is devoid of cross-term effects. Figure 2.11(b, c) show respectively the TFDS for $D = 3$ and $D = 6$. We see that at $D = 3$ it is possible to capture the most useful information in the time-frequency plane without the degrading effect of the cross terms.

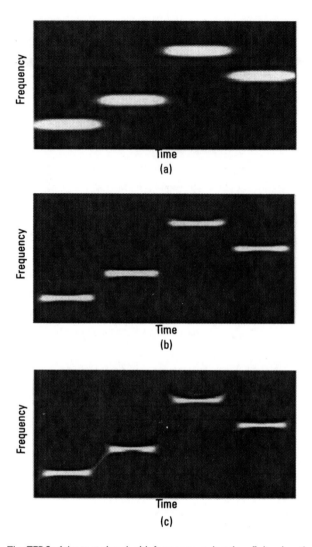

Figure 2.11 The TFDS of the test signal with four nonoverlapping, finite-duration sinusoids: (a) $D = 0$; (b) $D = 3$; and (c) $D = 6$.

In summary, we have described a number of popular time-frequency distributions in this chapter. The list includes the STFT, the CWT, the adaptive joint time-frequency representation, the WVD, Cohen's class, and the TFDS. These time-frequency transforms will be used in subsequent chapters of this book for various applications of radar imaging and signal analysis.

References

[1] Gabor, D., "Theory of Communication," *J. IEE (London)*, Vol. 93, No. III, November 1946, pp. 429–457.

[2] Cohen, L., *Time-Frequency Analysis*, Englewood Cliffs, NJ: Prentice Hall, 1995.

[3] Qian, S., and D. Chen, *Introduction to Joint Time-Frequency Analysis—Methods and Applications*, Englewood Cliffs, NJ: Prentice Hall, 1996.

[4] Moghaddar, A., and E. K. Walton, "Time-Frequency-Distribution Analysis of Scattering from Waveguide Cavities," *IEEE Trans. Antennas Propagat.*, Vol. AP-41, pp. 677–680, May 1993.

[5] Goupillaud, P., A. Grossman, and J. Morlet, "Cycle-Octave and Related Transforms in Seismic Signal Analysis," *Geoexploration*, Vol. 23, 1984, pp. 85–102.

[6] Heil, C. E., and D. F. Walnut, "Continuous and Discrete Wavelet Transforms," *SIAM Review*, Vol. 31, December 1989, pp. 628–666.

[7] Daubechies, I., "The Wavelet Transform, Time-Frequency Localization and Signal Analysis," *IEEE Trans. Inform. Theory*, Vol. IT-36, September 1990, pp. 961–1005.

[8] Combes, J. M., A. Grossmann, and P. Tchamitchian (eds.), *Wavelets, Time-Frequency Methods and Phase Spaces*, Berlin: Springer-Verlag, 1989.

[9] Qian, S., and D. Chen, "Signal Representation Using Adaptive Normalized Gaussian Functions," *Signal Processing*, Vol. 36, March 1994, pp. 1–11.

[10] Mallat, S. G., and Z. Zhang, "Matching Pursuits with Time-Frequency Dictionaries," *IEEE Trans. Signal Processing*, Vol. SP-41, December 1993, pp. 3397–3415.

[11] Trintinalia, L. C., and H. Ling, "Joint Time-Frequency ISAR Using Adaptive Processing," *IEEE Trans. Antennas Propagat.*, Vol. AP-45, February 1997, pp. 221–227.

[12] Qian, S., D. Chen, and Q. Yin, "Adaptive Chirplet Based Signal Approximation," *Proc. ICASSP*, Vol. III, May 1998, pp. 1871–1874.

[13] Bultan, A., "A Four-Parameter Atomic Decomposition of Chirplets," *IEEE Trans. Signal Processing*, Vol. 47, March 1999, pp. 731–745.

[14] Wigner, E. P., "On the Quantum Correction for Thermodynamic Equilibrium," *Phys. Rev.*, Vol. 40, 1932, pp. 749.

[15] Ville, J., "Theorie et Applications de la Notion de Signal Analytique," *Cables et Transmission*, Vol. 2, 1948, pp. 61–74.

[16] Page, C. H., "Instantaneous Power Spectra," *J. Appl. Phys.*, Vol. 23, 1952, pp. 103–106.

[17] Margenau, H., and R. N. Hill, "Correlation Between Measurements in Quantum Theory," *Prog. Theoret. Phys.*, Vol. 26, 1961, pp. 722–738.

[18] Rihaczek, A. W., "Signal Energy Distribution in Time and Frequency," *IEEE Trans. Inform. Theory*, Vol. IT-14, 1968, pp. 369–374.

[19] Cohen, L., "Generalized Phase-Space Distribution Functions," *J. Math. Phys.*, Vol. 7, 1966, pp. 781–806.

[20] Choi, H., and W. J. Williams, "Improved Time-Frequency Representation of Multicomponent Signals Using Exponential Kernels," *IEEE Trans. Acoustics, Speech, Signal Processing*, Vol. ASSP-37, June 1989, pp. 862–871.

[21]   Zhao, Y., L. E. Atlas, and R. J. Marks, "The Use of Cone-Shaped Kernels for Generalized Time-Frequency Representations of Nonstationary Signals," *IEEE Trans. Acoustics, Speech, Signal Processing*, Vol. ASSP-38, July 1990, pp. 1084–1091.

[22]   Qian, S., and D. Chen, "Decomposition of the Wigner-Ville Distribution and Time-Frequency Distribution Series," *IEEE Trans. Signal Processing*, Vol. SP-42, October 1994, pp. 2836–2842.

3

Detection and Extraction of Signal in Noise

The detection and extraction of an unknown signal in noise are important issues in radar signal processing. When a signal is severely corrupted by noise and cannot be observed in either the time domain or the frequency domain, a transformation whose basis functions are localized in both the time and the frequency domains, such as the Gabor transform, is very useful for observing the signal. By taking the time-frequency transform, random noise tends to spread its energy over the entire time-frequency domain, while signals often concentrate their energy within limited time intervals and frequency bands. Thus, signals embedded in noise are much easier to recognize in the joint time-frequency domain.

For the detection and extraction of weak signals in noise, we first need to detect those coefficients in the joint time-frequency domain that correspond to the desired signal. Then, we use only these coefficients to recover the time-domain signal waveform.

To detect the signal's coefficients, an appropriate threshold should be set up. If a coefficient is greater than the threshold, it is assigned to the signal. Otherwise, the coefficient is assigned to noise. An optimal way to set up the threshold is based on the constant false-alarm rate (CFAR) detection [1, 2]. In this chapter we extend the CFAR detection to the joint time-frequency domain. By setting a threshold with CFAR in the time-frequency domain, we can examine any coefficient to determine whether it belongs to a signal. Then, the signal can be extracted by taking the inverse time-

47

frequency transform using only the detected time-frequency coefficients. Thus, the signal buried in noise can be detected, its parameters can be measured, and its signal-to-noise ratio (SNR) can be enhanced.

In Section 3.1, we discuss how time-frequency transforms can be used for the detection and extraction of signals in noise. Then, in Section 3.2, we describe the concept of a time-varying frequency filter and demonstrate an example of the time-varying frequency filtering for a chirp signal. In Section 3.3, we analyze the SNR improvement by using time-varying frequency filters. Finally, in Sections 3.4 and 3.5, we discuss CFAR detection and extraction in the joint time-frequency domain.

3.1   Introduction

For signals corrupted by strong background noise, it is usually very difficult to perform signal detection and parameter estimation in either the time domain or the frequency domain. However, they may be identified in the joint time-frequency domain by taking a time-frequency transform as shown in Figure 3.1, where the time-frequency distribution series described in Chapter 2 is used. Signal often concentrates its energy within a limited time interval and a limited frequency band while random noise typically has energy spread over the time-frequency plane. Consequently, by representing the signal and noise in the joint time-frequency domain, signal detection becomes much easier. By applying time-varying frequency filtering, the SNR can also be enhanced. If we can distinguish those coefficients that belong to the signal from the ones that belong to the noise, these coefficients can be utilized to reconstruct the signal simply by taking the inverse time-frequency transform as shown in Figure 3.2.

To separate the signal coefficients from the noise coefficients, an appropriate threshold is needed. However, a fixed threshold is not suitable for detecting signals in different noise environments because the false-alarm rate may vary. Thus, an adaptive threshold that keeps a CFAR under various background noises is the most desirable. Human mental processes normally apply the CFAR function very well to distinguish useful signals from background noise and clutter. We will extend the CFAR detection to the joint time-frequency domain to extract unknown signals in background noise.

3.2   Time-Varying Frequency Filtering

To extract signals in a noisy environment, the conventional approach is to apply either frequency filtering or time gating. However, the frequency

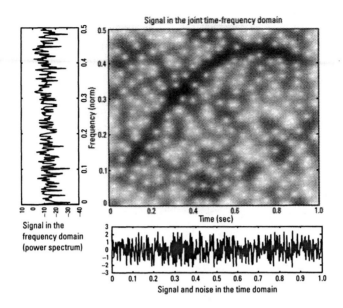

Figure 3.1 Signal may be identified in the joint time-frequency domain.

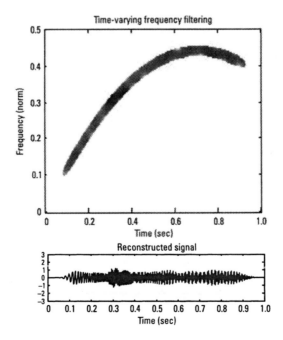

Figure 3.2 Time-varying frequency filtering and the reconstructed signal.

filtering will not remove noise within the frequency pass-band; and the time gating cannot remove noise within the gate. As illustrated in Figure 3.3, the time-varying frequency filter is the one that can remove the noise within its frequency pass-band and time-gate interval.

Unlike traditional linear frequency filtering that has a time-invariant frequency response, the time-varying frequency filter has a time-varying frequency response. Figure 3.4(a) shows a noisy signal in the time domain.

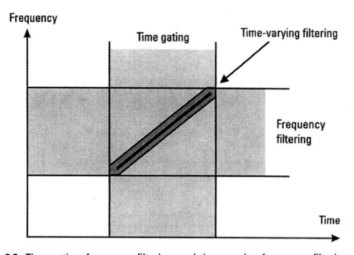

Figure 3.3 Time gating, frequency filtering, and time-varying frequency filtering.

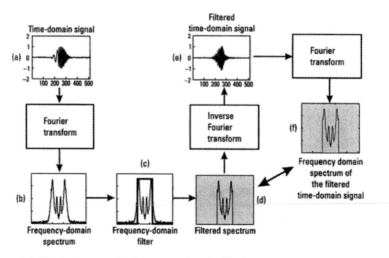

Figure 3.4 Block diagram of a frequency-domain filtering.

To apply a linear frequency filter to the noisy signal, we first take the Fourier transform of the noisy signal as shown in Figure 3.4(b), then multiply the coefficients of the Fourier transform by a desired frequency response function in Figure 3.4(c) and, finally, take the inverse Fourier transform to obtain a filtered signal as shown in Figure 3.4(e). Because the Fourier transform has a one-to-one mapping between the frequency domain and the time domain, any spectrum in the frequency domain corresponds to a unique signal in the time domain. Thus, the Fourier transform of the filtered signal in Figure 3.4(f) will have the desired frequency response as shown in Figure 3.3(d). However, in general, for time-frequency transforms there is no guarantee that such a one-to-one mapping exists between the time-frequency domain and the time domain. As illustrated in Figure 3.5, the above statement implies that the inverse time-frequency transform of the filtered signal shown in Figure 3.5(f) does not necessarily fall within the desired time-frequency region in Figure 3.5(d). Therefore, there is an issue concerning the time-frequency synthesis. Given a localized region in the time-frequency domain, how do we find the corresponding signal in the time domain? There are two approaches for solving this problem. The least-square-solution approach is to find the time-domain signal that minimizes the square error between the time-frequency transform of the signal and the desired time-frequency distribution [4–6]. Another approach proposed in [3] uses an iterative time-varying filtering as illustrated in Figure 3.6. First, it takes a time-frequency transform (such as the Gabor transform) of the noisy signal. Then, a desired time-varying frequency filter is applied to extract those coefficients that belong to the signal. By taking the inverse time-frequency transform, a time-frequency-filtered signal can be obtained. This completes the first iteration of the time-varying filter processing. Then, the same procedure can be applied to perform further iterations. It was proved in [7] that the first iteration of the time-varying filtering is exactly the least square solution and that further iterations will improve the least square solution.

3.3 SNR Improvement in the Time-Frequency Domain

As described in Chapter 1, the SNR is usually defined as the ratio of the average signal power to the average noise power. For a signal in the time domain, the average power is defined by

$$P_S = \frac{1}{T} \int_0^T s^2(t)dt \tag{3.1}$$

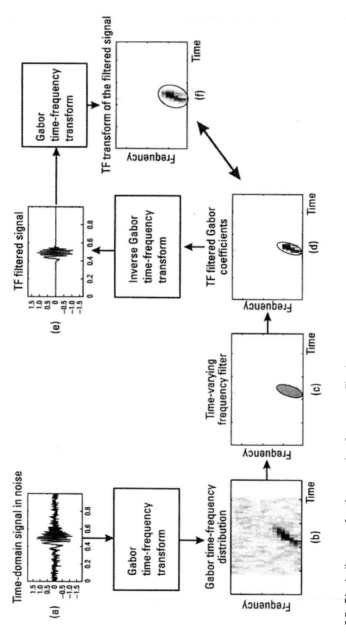

Figure 3.5 Block diagram of a time-varying frequency filtering.

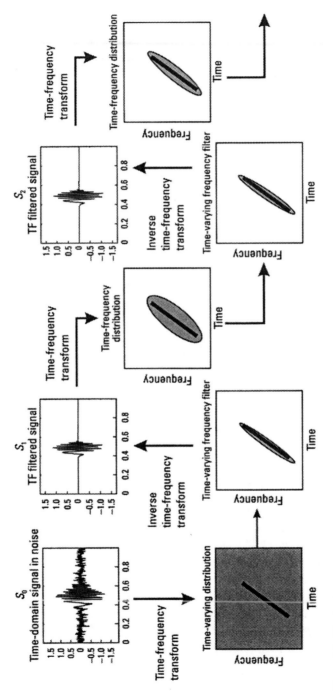

Figure 3.6 Block diagram of an iterative time-varying frequency filtering.

and the average noise power is defined by

$$P_N = \int_{-\infty}^{\infty} (r_n - mean\{r_n\})^2 p(r_n) dr_n \qquad (3.2)$$

where $r_n = \{n(t)\}$ is a random process described by an ensemble of the random noise function, and $p(r_n)$ is the probability density function of the random process r_n. Thus, the SNR becomes

$$SNR_{average} = \frac{P_S}{P_N} \qquad (3.3)$$

For an additive white Gaussian noise with zero-mean and variance $\sigma_{r_n}^2$, the average noise power is $P_N = \sigma_{r_n}^2$, and the SNR becomes

$$SNR_{average} = \frac{\frac{1}{T}\int_0^T s^2(t)dt}{\sigma_{r_n}^2} \qquad (3.4)$$

According to this SNR definition, an orthogonal transformation such as the Fourier transform does not change the SNR. Thus, by taking the Fourier transform the SNR in the frequency domain is equal to the SNR in the time domain.

3.3.1 SNR Definition Suitable for Signal Detection and Extraction

Let us examine a sinusoidal signal buried in additive white Gaussian noise with zero-mean and variance $\sigma_{r_n}^2$ as shown in Figure 3.7(a). According to the SNR defined by (1.15), because the average signal power is very low, the SNR is so low that it is impossible to distinguish the signal from noise in the time domain. However, if we take the Fourier transform, a signal peak can be seen clearly in the frequency domain as shown in Figure 3.7(b). This suggests that the SNR definition must be modified for signal detection and extraction. For this purpose, we are not interested in the average signal power. Instead, we are interested in the peak power of the signal. The peak-power SNR can be defined by modifying the instantaneous-power SNR in (1.17) with the peak signal power:

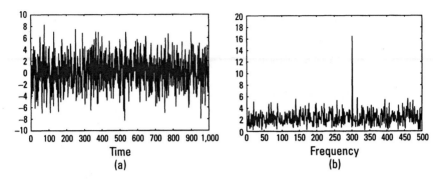

Figure 3.7 A sinusoidal signal in noise. (Data provided by X. G. Xia.)

$$SNR_{peak} = \frac{P_{peak}}{\sigma_{r_n}^2} \qquad (3.5)$$

where $P_{peak} = \max_{0 < t < T} \{s^2(t)\}$. Thus, the peak-power SNR defined here can be used for evaluating system performance in signal detection.

Another similar SNR definition proposed in [8] can also be used for signal detection. Assuming the peak signal power is P_{peak}, we want to find a region ΔT where the instantaneous signal power is above the half-power points (i.e., the points at which the power is one-half of P_{peak}, or 3 dB below, the peak power):

$$\Delta T = \{t : 0 < t < T \ and \ |s^2(t)| \geq 0.5 P_{peak}\} \qquad (3.6)$$

and the SNR can be defined as

$$SNR_{\Delta T} = \frac{\frac{1}{\Delta T} \int_{t \in \Delta T} s^2(t) dt}{\sigma_{r_n}^2} \qquad (3.7)$$

With this definition, the SNR of the noisy signal in the time domain shown in Figure 3.7(a) is −8.4 dB. Because the SNR definition is suitable to signals in any transform domain, we can also use the same SNR definition to analyze the same signal but in the frequency domain. Thus, the SNR turns out to be 16.3 dB. Therefore, the SNR improvement is about 24 dB in the frequency domain. This is the reason why the sinusoidal signal buried in additive white Gaussian noise with zero-mean and variance $\sigma_{r_n}^2$ as shown in Figure 3.7(a) can be easily detected in the frequency domain.

3.3.2 SNR in the Joint Time-Frequency Domain

As mentioned earlier in this chapter, by taking a time-frequency transform, the noise tends to spread its energy over the time-frequency domain, while the signal often concentrates its energy into regions within limited time intervals and frequency bands. Especially for a frequency-modulated signal, such as $\exp\{j2\pi[f_0 + (\eta/2)t]t\}$ embedded in noise, it is much easier to be recognized in the joint time-frequency domain than in either the time or the frequency domain alone. However, when the frequency-changing rate η becomes very large, to keep the same signal energy the signal becomes a short time impulse that can be easier recognized in the time domain; when η approaches zero, the signal becomes a sinusoid $\exp\{j2\pi f_0 t\}$ that can be more easily recognized in the frequency domain. In general, there is a question about how much the SNR can be improved with the time-frequency transform? The answer is that it depends on the type of the time-frequency transform and the waveform of the signal. In [8], a quantitative analysis of SNR of a multicomponent signal with the STFT is given. The multicomponent signal consists of a number of monocomponent signals

$$s(t) = \sum_{k=1}^{K} s_k(t) \tag{3.8}$$

Because the time-frequency transform is especially suitable to represent signals with time-varying spectrum, the monocomponent signal is assumed to be a chirp-type signal

$$s_k(t) = a_k(t) \exp\left\{j2\pi\left(f_k + \frac{\eta_k}{2}t\right)t\right\} \tag{3.9}$$

where $a_k(t)$ is the amplitude function of the kth component, f_k is the starting frequency of the kth chirp signal, and η_k is the chirp rate of the kth component.

It was proved in [8] that for the multicomponent signal in additive Gaussian white noise with zero-mean and variance $\sigma_{r_n}^2$, the SNR improvement using the STFT with a rectangular window over the SNR in the time domain is equal to or greater than the order of (N/K):

$$\frac{SNR_{STFT}}{SNR_T} \geq O\left(\frac{N}{K}\right) \tag{3.10}$$

where N is the number of the samples within the short time window, and K is the number of monocomponents in the multicomponent signal. In [9], it was proved that the SNR improvement in the joint time-frequency domain using the pseudo WVD over the SNR in the time domain is

$$\frac{SNR_{PWV}}{SNR_T} \geq O\left(\frac{N}{K^2}\right) \tag{3.11}$$

To improve the SNR in the time-frequency domain, higher sampling rate and smaller number of monocomponents are desirable.

When the chirp rate in the chirp-type monocomponent signal is zero, the chirp signal becomes a sinusoidal signal. In this case, the SNR in the joint time-frequency domain will be equal to the SNR in the frequency domain. If the chirp rate becomes very large, the chirp signal becomes an impulse signal and the SNR in the time-frequency domain is equal to that in the time domain. In order to gain a SNR improvement in the time-frequency domain, the chirp rate must be within a certain range and must be neither too large nor too small. The bounds have been found in [10] to be

$$1/(0.8\sqrt{2\pi})^2 < \max_{1 \leq k \leq K} |\eta_k| < 1.28\pi \tag{3.12}$$

3.4 CFAR Detection in the Joint Time-Frequency Domain

An observed waveform in the radar receiver can be either "a signal corrupted by noise," or "noise alone." The objective of signal detection is to decide whether or not there is a signal present in the observation, subject to a certain false-alarm rate. When a signal is detected in the observation, parameters of the signal, such as time delay, time duration, center frequency, and frequency change rate, can also be estimated by extracting the signal from the observed waveform.

Classical signal detection with a fixed threshold as shown in Figure 3.8, where probability distribution functions of the signal and noise are given, is not suitable for detecting unknown signals in a statistically nonstationary

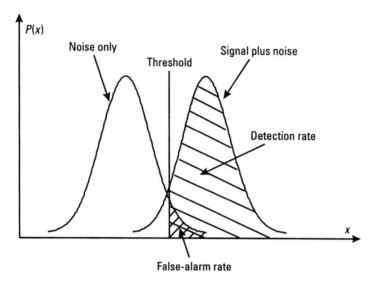

Figure 3.8 Classical detection of signal in noise.

environment because the false-alarm rate varies. Therefore, an adaptive CFAR threshold is very useful under different background noise environments. The adaptive CFAR method [1, 2] estimates the statistical characteristics of the background noise or clutter from the neighborhoods of a test cell and, then, sets a detection threshold based on the estimation to determine whether the test cell belongs to the signal as illustrated in Figure 3.9.

In many practical situations, in the joint time-frequency domain the statistical amplitude distribution $p(x)$ of the background noise in the time-frequency domain can be well approximated with a Rayleigh distribution shown in Figure 3.10, and expressed as

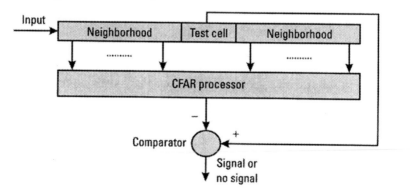

Figure 3.9 CFAR detection of a signal in noise.

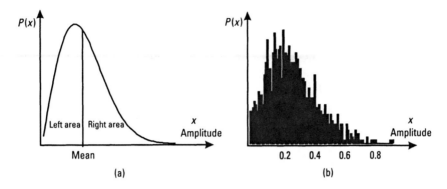

Figure 3.10 (a) Rayleigh distribution, and (b) distribution of an observed noise after envelope detector in the joint time-frequency domain.

$$p(x) = \frac{x}{\mu^2} \exp\left\{-\frac{x^2}{2\mu^2}\right\} \tag{3.13}$$

where μ is related to the mean value by

$$mean = \sqrt{\frac{\pi}{2}}\,\mu \tag{3.14}$$

The left-side area of the mean is equal to the right-side area, $A\_left = A\_right$, as indicated in Figure 3.10(a).

Since the amplitude distribution of the background noise is approximately a Rayleigh distribution, the CFAR threshold can be determined with respect to the Rayleigh amplitude distribution [11]. Thus, the background noise in the joint time-frequency domain would be detected with the CFAR. The number of missed points in the time-frequency domain depends on the required false-alarm rate. For those points that are below the CFAR threshold, they would be considered as the background noise, while all other points above the CFAR threshold would be considered as the signal plus the background noise.

The false-alarm rate should be chosen low enough so that not a great number of background points are mistaken for signals. The choice of the false-alarm rate depends on the nature of the signal expected. The lower the false-alarm rate, the larger the number of weak signal points will be mistaken for background noise; the higher the false-alarm rate, the larger the number of strong noise points that will be mistaken for signals. Therefore, an appropriate false-alarm rate should be selected such that no significant number of noise points is mistaken for signal.

To determine the CFAR threshold, we should estimate the mean of the Rayleigh distribution of the observed background noise. Usually, the distribution of the background noise can be obtained by calculating the amplitude distribution of the observed waveform over the noise-only regions in the joint time-frequency domain. Figure 3.10(b) shows the distribution of an observed waveform after envelope detector in the radar receiver. It obeys the Rayleigh distribution with the mean value of 0.26 for this observed specific waveform.

Assume the observed waveform is

$$x(t) = s(t) + n(t) \qquad (3.15)$$

then, the CFAR detection can be described as

$$\frac{x(t)}{\mu} = \begin{cases} > Th & (\text{for signal}) \\ \leq Th & (\text{for noise}) \end{cases} \qquad (3.16)$$

where Th is a threshold determined by [11]

$$Th = \sqrt{2\log_e\left(\frac{1}{P_{fa}}\right)} \qquad (3.17)$$

where the false-alarm rate P_{fa} is usually set to be 10^{-4}–10^{-5}.

The procedure for detecting unknown signals in noise is illustrated in Figure 3.11 and described as follows [12]:

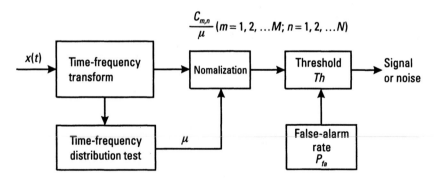

Figure 3.11 Block diagram of the CFAR detection.

1. Compute the time-frequency coefficients of the observed waveform $x(t)$:

$$C_{m,n} = \langle x(k), g_{m,n}(k)\rangle_k \ (m = 1, 2, \ldots M; \ n = 1, 2, \ldots N)$$

where k is the index of time sampling, $g_{m,n}(k)$ is a time-frequency basis function such as a Gabor function, $\langle x(k), g_{m,n}(k)\rangle_k$ is the inner product of $x(k)$ with $g_{m,n}(k)$ over k that gives the time-frequency coefficient of the waveform $x(k)$, and $C_{m,n}$ is the $M \times N$ time-frequency coefficient, where M is the number of time samples and N is the number of frequency samples.

2. Determine the mean value of the Rayleigh distribution μ from the noise-only region in the joint time-frequency domain.

3. Calculate the CFAR threshold Th based on a given false-alarm rate P_{fa}.

4. Determine those time-frequency coefficients whose value are above the threshold and set the rest of the time-frequency coefficients to zero.

3.5 Signal Extraction in the Joint Time-Frequency Domain

In the joint time-frequency domain noise tends to spread its energy over the time-frequency domain, while signals often concentrate their energy on regions with limited time intervals and frequency bands. Thus, with CFAR detection, signals can be detected and reconstructed by using the detected time-frequency coefficients [12].

3.5.1 Time-Frequency Expansion and Reconstruction

The first step for reconstructing unknown signals is to eliminate the noise in the joint time-frequency domain and obtain the noise-suppressed time-frequency coefficients. Then, the second step is to apply the inverse time-frequency transform to reconstruct the signal waveform.

It is known that if a linear time-frequency transform, such as the Gabor transform, is taken for a signal, the signal can be perfectly reconstructed from its time-frequency coefficients [13]. However, this is not always true if the time-frequency coefficients are modified by a 2D mask function. In general, the modified coefficients are not valid time-frequency coefficients.

The reconstruction based on the modified coefficients may not lead to the desired signal waveform. Therefore, some kind of criterion is needed to estimate the anticipated signals. The most commonly used criterion is the least squares solution method mentioned in Section 3.2, which is to find a signal waveform whose time-frequency coefficients are close to the desired coefficients in the least squares sense, that is:

$$\xi = \min_{\hat{x}} \sum_{m=0}^{M-1} \sum_{n=0}^{N-1} \left| \tilde{C}_{m,n} - <\hat{x}(k), g_{m,n}(k)>_k \right|^2 \qquad (3.18)$$

where k is the index of time sampling, $\hat{x}(k)$ denotes the estimated waveform, $g_{m,n}(k)$ is a time-frequency basis function, the inner product $<\hat{x}(k),$ $g_{m,n}(k)>_k$ gives the time-frequency coefficient of the estimated signal, and $\tilde{C}_{m,n}$ is the desired $M \times N$ time-frequency coefficient, where M is the number of time samples and N is the number of frequency samples.

3.5.2 Time-Frequency Masking and Signal Extraction

By applying a 2D mask function to the time-frequency coefficients $C_{m,n}$, the modified time-frequency coefficients are as the follows

$$\tilde{C}_{m,n} = \begin{cases} C_{m,n} & \text{if } M_{m,n} = 1 \\ 0 & \text{if } M_{m,n} = 0 \end{cases} \qquad (3.19)$$

Figure 3.12(a) shows an example of an observed waveform that consists of an unknown signal and noise at a SNR of 0 dB. The Gabor time-frequency coefficients of the observed waveform in gray scales are shown in Figure 3.12(b) and the modified coefficients by applying the mask function are in Figure 3.12(d). The mask function is obtained by applying the CFAR detection. The extracted signal waveform whose time-frequency coefficients are close to the desired coefficients in (3.19) in the least squares sense. Compared with the observed waveform in Figure 3.12(a), the extracted signal in Figure 3.12(c) is a noise-removed chirp waveform.

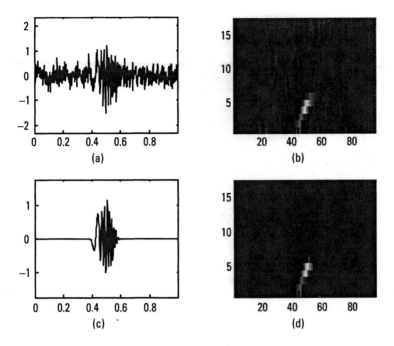

Figure 3.12 (a) A signal in noise at an SNR of 0 dB; (b) Gabor time-frequency distribution of the observed signal in noise; (c) the extracted signal; and (d) Gabor time-frequency coefficients used to construct the signal.

References

[1]   Skolnik, M. I., *Radar Handbook, Second Edition,* New York: McGraw-Hill, 1990.

[2]   Barton, D. K., *Modern Radar System Analysis,* Norwood, MA: Artech House, 1988.

[3]   Qian, S., and D. Chen, *Joint Time-Frequency Analysis—Methods and Applications,* New Jersey: Prentice Hall, 1996.

[4]   Boudreaux-Bartels, G. F., and T. W. Parks, "Time-Varying Filtering and Signal Estimation Using Wigner Distribution Synthesis Techniques," *IEEE Trans. on Acoust., Speech, Signal Processing,* Vol. 34, No. 6, 1986, pp. 442–451.

[5]   Farkash, S., and S. Raz, "Time-Variant Filtering Via the Gabor Expansion," *Signal Processing V: Theories and Applications,* New York: Elsevier, 1990, pp. 509–512.

[6]   Hlawatsch, F., and W. Krattenthaler, "Time-Frequency Projection Filters and Time-Frequency Signal Expansion," *IEEE Trans. on Signal Processing,* Vol. 42, No. 12, 1994, pp. 3321–3334.

[7]   Xia, X. G., and S. Qian, "Convergence of an Iterative Time-Variant Filtering Based on Discrete Gabor Transform," *IEEE Trans. on Signal Processing,* Vol. 47, No. 10, 1999, pp. 2894–2899.

[8]   Xia, X. G., "A Quantitative Analysis of SNR in the Short-Time Fourier Transform Domain for Multicomponent Signals," *IEEE Trans. on Signal Processing*, Vol. 46, No. 1, 1998, pp. 200–203.

[9]   Xia, X. G., and V. C. Chen, "A Quantitative SNR Analysis for the Pseudo Wigner-Ville Distribution," *IEEE Trans. on Signal Processing*, Vol. 47, No. 10, 1999, pp. 2891–2894.

[10] Xia, X. G., G. Y. Wang, and V. C. Chen, "A Quantitative Signal-to-Noise Ratio Analysis for ISAR Imaging Using Joint Time-Frequency Analysis—Short Time Fourier Transform," *IEEE Trans. on Aerospace and Electronic Systems*, 2000.

[11] Lank, G. W., and N. M. Chung, "CFAR for Homogeneous Part of High-Resolution Imagery," *IEEE Trans. on Aerospace and Electronic Systems*, Vol. 28, No. 2, 1992, pp. 370–381.

[12] Chen, V. C., and S. Qian, "CFAR Detection and Extraction of Unknown Signal in Noise with Time-Frequency Gabor Transform," *SPIE Proc. on Wavelet Applications*, Vol. 2762, 1996, pp. 285–294.

[13] Qian, S., and D. Chen, "Discrete Gabor Transform," *IEEE Trans. on Signal Processing*, Vol. 41, No. 7, 1993, pp. 2429–2439.

4

Time-Frequency Analysis of Radar Range Profiles

Radar is an instrument traditionally used to pinpoint the position and velocity of a target from its back-scattered microwave energy. The development of high-resolution radar techniques in the past three decades has led to much more advanced radar capabilities in gathering information on the fine features of a target in addition to its position and velocity [1]. For instance, by using sufficient frequency bandwidth, it is possible to generate a 1D down-range map of the target called the range profile. Similarly, by observing a target in relative motion with respect to the radar over a sufficient time interval, it is possible to generate a 1D cross-range map of the target.

A range profile is basically a time history of the radar back-scattered signal due to a short pulse. Since time delay τ is related to the distance R along the radar LOS via the relationship $\tau = 2R/c$, where c is the speed of electromagnetic wave propagation, the resulting radar signal as a function of time can be interpreted as a mapping of the reflectivity of the target along the radar LOS, or the down-range direction. In simple targets, a range profile typically consists of a series of distinct peaks that can be related spatially to the isolated scattering centers on the target. These features are often utilized for signature diagnostic and target recognition applications. In real targets, however, the scattering physics is usually more complex. For example, the scattering from some components on a target is not always well-localized in time and may give rise to range-extended returns. The interpretation of these dispersive scattering phenomena is not as easy to carry out from the time-domain range profile.

In this chapter, we examine the use of time-frequency analysis for analyzing radar range profiles. In Section 4.1, we briefly review radar scattering phenomenology from a fundamental electromagnetics perspective (viz., both the theoretical foundation that leads to the well-known point-scatterer model and higher-order scattering physics that deviates from the simple point-scatterer model). In Section 4.2, we introduce the time-frequency analysis of range profiles and show how complex electromagnetic scattering mechanisms can be revealed in the joint time-frequency space. In Section 4.3, we illustrate the use of high-resolution techniques discussed in Chapter 2 for localizing and extracting the time-frequency scattering features. In Section 4.4, we demonstrate the extension of time-frequency processing to 2D radar imagery for extracting and interpreting complex scattering phenomena.

4.1 Electromagnetic Phenomenology Embedded in Back-Scattered Data

It is well known that radar targets, especially man-made targets, can often be considered as a collection of discrete point-scatterers. This model, called the point-scatterer model or the scattering center model, is widely used in many radar applications. Figure 4.1 illustrates the conceptual idea of this model, where the electromagnetic back-scattered signal from a complex target

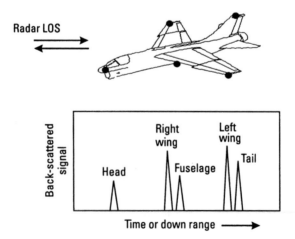

Figure 4.1 Point-scatterer model for a complex target and the associated high-resolution range profile.

can be thought of as if it is emanating from a set of scattering centers on the target. As a result, the high-resolution range profile becomes a 1D mapping of the geometrical point features on the target.

The point-scatterer model can be rigorously related to first-principle electromagnetic scattering theory through high-frequency ray optics, or the so-called GTD [2, 3]. In GTD, the scattering of an electromagnetic wave from a complex target at high frequencies is described by a set of highly localized ray phenomena, which are each attributable to a reflection or diffraction point on the target. As shown in Figure 4.2, these points can include specular reflections from smooth surfaces, edge diffractions from edges and tips, as well as multiple scattering from dihedral and trihedral corner reflectors. These points correspond exactly to the scattering centers in the point-scatterer model. Based on GTD, the total back-scattered field due to a monochromatic incident wave (with angular frequency $\omega = 2\pi f$ and time dependence $\exp\{j\omega t\}$) can then be written as

$$E^s(\omega) = \sum_n A_n \exp\{-j\omega(2R_n/c)\} \qquad (4.1)$$

where R_n is the down-range location of a scattering center along the radar LOS, and A_n is the scattering amplitude of the scattering center.

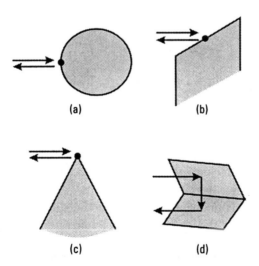

Figure 4.2 Ray optical descriptions of various scattering mechanisms: (a) surface reflection, (b) edge diffraction, (c) cone tip diffraction, and (d) dihedral corner reflection.

Several comments are in order. First, GTD is a "high-frequency" approximation to the rigorous Maxwell's equations. It is an appropriate approximation provided that the size of the target is large compared to the radar wavelength. For example, if an X-band radar operating at 3 cm wavelength is used to illuminate an air target of 20m size, the size of the target is 667 wavelengths and falls well within the ray optical regime where GTD is valid. Second, the scattering amplitudes of a number of canonical configurations have been derived over the years by electromagnetics researchers. For example, the diffraction amplitude from a conducting wedge is known in closed form [4, 5]. However, for more general structures, the scattering strengths are not always available analytically. Nevertheless, the basic idea of (4.1) (i.e., the total scattering can be written as a sum of the contributions from individual scattering centers), is well accepted even for more complex radar targets.

Next, we consider the case when the radar data is collected over a band of frequencies with bandwidth $\Delta\omega$ centered about $\omega_0 = 2\pi f_0$. If we assume the scattering amplitudes for all the scattering centers are independent of frequency, the time-domain scattered field (or the range profile) can be written as

$$E^s(R) = \sum_n A_n h\left(\frac{\Delta\omega}{c}(R - R_n)\right) \tag{4.2a}$$

where

$$h(\cdot) = \exp\{j(2\omega_0/\Delta\omega)(\cdot)\}\,\mathrm{sinc}(\cdot) \tag{4.2b}$$

is usually called the "point spread response" function of the scattering center. From the argument of the sinc function, we can see that the wider the radar bandwidth, the more focused h is in range. Therefore, given sufficient bandwidth, the radar range profile can be used to map out the different scattering centers on the target, as was illustrated in Figure 4.1.

We now turn our attention to discuss how real scattering phenomenology can deviate from the idealized point-scatterer model. The first deviation from the model in (4.1) is that the scattering amplitudes A_n are in general frequency dependent. Based on GTD, the scattering amplitudes of canonical conducting shapes have been shown to have an ω^{γ_n} dependence where γ_n can take on half-integer values such as -1 (corner), $-1/2$ (edge), 0 (doubly curved surface), $+1/2$ (singly curved surface) and $+1$ (dihedral or flat plate at specular) [6]. As a result of the frequency dependence, the point spread response is in general more spread out in range than the frequency independent case. For example, for $\gamma_n = 1$ the corresponding point spread response involves the derivative of the sinc function (since multiplication by ω translates into differentiation in time), which has a slightly larger width in range than the $\gamma_n = 0$ response. However, this type of deviation from the idealized point-scatterer model is quite small, especially if the bandwidth of the data is not very large.

The second type of deviation arises if a scattering mechanism involves ray propagation through a frequency dispersive region. In this case, the phase of the model in (4.1) should account for not only the ray propagation path in free space, but also that in the dispersive region. Denoting the path lengths over each region as R_{n1} and R_{n2}, we must modify the phase as

$$\phi_n(\omega) = -2\left[\frac{\omega}{c}R_{n1} + \beta(\omega)R_{n2}\right] \qquad (4.3)$$

The behavior of $\beta(\omega)$ as a function of frequency is dictated by the detailed dispersive characteristics of the propagation medium. If $\beta(\omega)$ is not a linear function of frequency, the resulting range behavior of the scattering mechanism can be more complex. In general, this means a well-localized incident pulse will become much more spread out in range after the scattering process.

Finally, there are situations when the scattering response completely deviates from the point-scatterer model. Typically, this occurs when the incident wavelength is on the order of the dimension of a scattering structure. Under this situation, the scattering mechanisms deviate significantly from the ray-optical description of the scattering process. In this so-called "resonant region," the response from a target feature can be very large at certain frequencies. Physically, we can think of this phenomenon as a strong constructive interference of the many multiple scattering mechanisms existing within the structure. This type of high-Q, resonant response in frequency usually translates into extended ringing in the range dimension. Interpretation of these range-extended returns is difficult, as they no longer convey the geometrical information that the scattering centers carry. In the next section, we introduce joint time-frequency representations that can display the scattering center information while simultaneously capturing the frequency dispersion and resonance information in the back-scattered signal.

4.2 Time-Frequency Representation of Range Profiles

The usefulness of the joint time-frequency analysis of signals has long been recognized in the signal processing arena [7, 8]. Recently, joint time-frequency methods have been applied to electromagnetic scattering data with good success [9–12]. The joint time-frequency representation of a signal is a 2D phase space that facilitates the visualization and interpretation of complex electromagnetic wave phenomenology. In this feature space, discrete time events such as scattering centers, discrete frequency events such as resonances, and dispersive mechanisms due to surface waves and guided modes can be simultaneously displayed. This can oftentimes lead to more insights into the complex electromagnetic wave propagation and scattering mechanisms than what is available in the traditional time or frequency domain alone.

The most common tool in generating the joint time-frequency representation of a time signal is the STFT introduced in Chapter 2. The STFT of the range profile $E^s(R)$ [denoted by $s(t)$ where $t = 2R/c$] is

$$STFT(t, \omega) = \int s(t')w(t' - t) \exp\{-j\omega t'\}dt' \qquad (4.4)$$

where $w(\cdot)$ is a short-time window function. The resulting 2D magnitude display of $|STFT(t, \omega)|$ is called the spectrogram. The spectrogram provides information on the frequency content of the signal at different time instances. Shown in Figure 4.3 are the time-frequency features of some commonly encountered scattering mechanisms discussed in the last section. A discrete event in time is due to wave scattering from a spatially localized scattering center on a structure. It shows up as a vertical line [Figure 4.3(a)] in the image since it occurs at a particular time instance but over all frequencies. A target resonance (e.g., the return from a partially open cavity), is a scattering event that becomes prominent at a particular frequency. It shows up as a horizontal line in the joint time-frequency plane [Figure 4.3(b)]. Dispersive phenomena, on the other hand, are characterized by slanted curves in the time-frequency image. For instance, surface wave mechanisms due to material coatings are characterized by curves with a positive slope [Figure 4.3(c)]. Another type of dispersion arises from waveguide structures. These structural dispersion mechanisms are characterized by curves with a negative slope in the time-frequency image [Figure 4.3(d)]. The detailed behavior of the slants is dependent on how the propagation velocity of the wave varies as a function of frequency. All of the above mentioned phenomena have been observed in a wide variety of structures, from simulation data on canonical structures

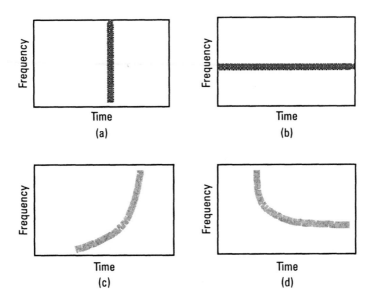

Figure 4.3 Electromagnetic mechanisms are manifested in the joint time-frequency image as distinct features: (a) scattering center; (b) resonance; (c) material dispersion; and (d) structure dispersion. (*Source:* [13] © 1998 SPIE.)

to measurement data on complex platforms. Below, several examples are presented to demonstrate the unique features of electromagnetic scattering mechanisms in the joint time-frequency plane.

In the first example, we consider a perfectly conducting strip containing a resonant cavity near the middle of the strip shown in Figure 4.4(a). A small fin exists at the right edge of the strip. Although this target is simple, its scattering is representative of signatures from more realistic targets with both exterior skinline contributions and subskinline resonances. For a radar wave with its electric field polarized perpendicularly to the plane of the paper and incident at 25 degrees from edge-on, three scattering centers due to the left edge, the cavity exterior, and the fin at the right edge should arise in the range profile. In addition, six cavity resonances are expected to be excited in the frequency range 0.5 to 18 GHz. Figure 4.4(b) shows the frequency-domain scattered far-field generated by a 2D moment method code based on an electric field integral equation formulation. Although the cavity resonances are expected to appear as sharp spikes in the frequency domain, they are overshadowed by the frequency behavior of the three scattering centers on the strip. Consequently, it is difficult to distinguish the resonances in Figure 4.4(b). The time-domain data, shown in Figure 4.4(c), was obtained

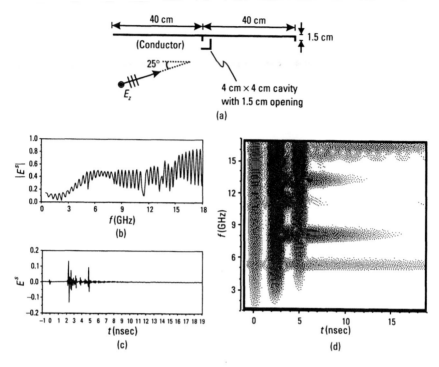

Figure 4.4 Back-scattered data from a conducting strip with an open cavity: (a) geometry of the target; (b) frequency domain data; (c) time domain data; and (d) spectrogram obtained via the STFT. (*Source:* [14] © 1995 IEEE.)

by inverse-Fourier transforming the 0.5 to 18 GHz data. A Kaiser window was applied to the frequency data before the transform. The locations of the three scattering centers (at t ~ 0, 2, and 5 nsec) can be seen in Figure 4.4(c), but the time-behavior of the cavity resonances makes it difficult to resolve each scattering center clearly. This is particularly true for the third scattering center, which is excited before the resonances have significantly decayed. Shown in Figure 4.4(d) is the spectrogram of the scattered data obtained via the STFT. The three vertical lines correspond to the scattering centers, and the five visible horizontal lines correspond to the cavity resonances. As we can see, both scattering center phenomenon and the weaker (although still interesting) resonances can be simultaneously displayed to give a good picture of all the key scattering features on this target.

In the second example, the back-scattered data from a dielectric-coated plate with a gap in the coating is considered [15]. The structure is shown in the upper left corner of Figure 4.5. The radar signal is incident edge-on from the left with the incident electric field polarized in the vertical direction.

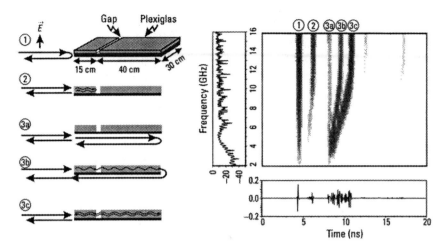

Figure 4.5 Joint time-frequency image of the back-scattering data from a coated plate with a gap in the coating. The JTF image is generated by the STFT. Those features which show slanting in the JTF plane are associated with the dispersive surface wave mechanisms in the coating. (*Source:* [15] © 1993 IEEE.)

The radar frequency response was generated by computer simulation using the method of moments and verified by laboratory measurement. The simulation result is shown along the vertical frequency axis. The time-domain response, or equivalently the range profile, was obtained by inverse-Fourier transforming the 1.7-to-18-GHz data. The resulting response is shown along the horizontal time axis. It appears that three distinct pulses are present. However, the second and third pulses are spread out in range. In order to resolve in finer detail the dispersive scattering mechanisms in this coated plate, the spectrogram of the back-scattered signal is generated using the STFT. As can be seen, the scattering mechanisms are much more apparent in the 2D joint time-frequency plane than in either the time or the frequency domain. In particular, it is observed that the third broad pulse in the time domain actually consists of three separate scattering mechanisms (labeled 3a, 3b, and 3c). As the frequency approaches zero, the propagation delays of mechanisms 3a, 3b, and 3c approach the same value. As frequency increases, the pulses have different propagation delays and become clearly separated. Slanted curves in the time-frequency plane (like mechanisms 2, 3b, and 3c) are characteristic of dispersive behavior. In the case of the coated plate, surface waves excited in the coating give rise to the dispersive mechanisms. At frequencies well above cutoff, the surface wave is tightly bound to the dielectric and the wave velocity approaches the slower dielectric velocity.

Near cutoff, the surface wave velocity approaches that of free space and exhibits a shorter propagation delay. Therefore, in the time-frequency plane, surface wave phenomena show up as slanted curves with a positive slope. Based on propagation delay considerations and the above observation, it is possible to pinpoint the five dominant scattering mechanisms. They are shown on the left in Figure 4.5, which clearly indicates that mechanisms 2, 3b, and 3c include surface wave propagation. For the polarization under consideration and the frequency range of the data, the TM_0 surface wave mode, which has zero cutoff, is the only mode that can propagate in the dielectric (the TM_1 mode has a cutoff frequency of 23.4 GHz). Finally, higher order scattering mechanisms can be observed during the late-time portion of the data but are very weak.

In the third example, we consider the scattering from a conductor-backed dielectric grating shown in Figure 4.6 [16]. The finite grating contains twelve triangular grooves of equal width. An incident plane wave with its magnetic field polarized perpendicularly to the plane of the paper is incident at an angle of 30 degrees from edge-on. The back-scattered frequency data from the grating was generated by computer simulation using a 2D method of moments code. The frequency domain response is shown along the vertical

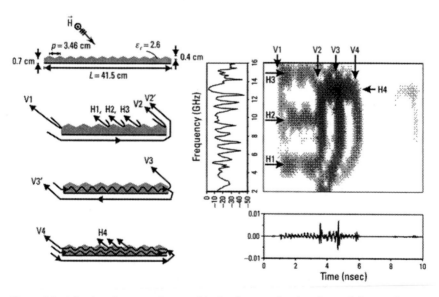

Figure 4.6 Joint time-frequency image of the back-scattering data from a finite, conductor-backed periodic grating. The JTF image is generated by the STFT. The dominant scattering mechanisms are labeled V1–V4 and H1–H4. (*Source:* [16] © 1996 Wiley.)

axis and the corresponding time response is shown along the horizontal axis of Figure 4.6. Based on infinite grating theory, frequency resonances corresponding to the retro-reflection (or back-scattering) of Floquet harmonics are expected to occur at 5, 10, and 15 GHz. As can be seen from the frequency spectrum, the two most pronounced resonances occur at 5 and 13.1 GHz. The resonances expected at 10 and 15 GHz are extremely difficult to distinguish from the background signal. In addition, the 13.1-GHz peak is rather surprising because it cannot be explained in terms of a Floquet mode excited by the incident plane wave. From the time-domain perspective, a number of distinct, yet broad, pulses are visible. The broadness of the pulses is due to dispersive surface wave mechanisms present in the dielectric grating and the resonant nature of the Floquet harmonics being excited. In order to better resolve the dispersive scattering mechanisms and resonances present in the coated strip, the STFT is used to generate the spectrogram in Figure 4.6. Present in the spectrogram are four horizontal lines (labeled H1–H4) and four vertically oriented lines (labeled V1–V4). These eight features are related to the scattering mechanisms labeled in the illustrations on the left in Figure 4.6. The fact that the spectrogram shows such distinct features makes it easy to pinpoint the various mechanisms, including the Floquet harmonics due to the incident wave, the Floquet harmonics due to surface waves, and the diffraction mechanisms due to the edges of the grating. The slanted behavior of V3 and V4 is due to the dispersive behavior of surface waves supported by the dielectric grating and has been explained in the last example. H2 and H3, which were barely visible as resonances in the frequency spectrum, are easy to distinguish when viewed in the time-frequency plane. Moreover, mechanism H4 is caused by the Floquet harmonics due to the surface waves excited by the edges. It is a scattering mechanism that is unique to finite periodic structures.

In the fourth example, the scattering from a slotted waveguide structure is considered [17]. The geometry is shown in Figure 4.7, where a long rectangular waveguide is flush mounted in a conducting ground plane. Two narrow slots are opened on each end of the ground plane. The structure is excited by a plane wave with a horizontally polarized electric field at an angle of 30 degrees with respect to the vertical. The back-scattered data was generated by computer simulation based on the method of moments from 0.025 to 10 GHz. The spectrogram, obtained using the STFT, is shown on the right in Figure 4.7. The two early-time vertical lines correspond with the exterior scattering centers from the slots. The other curves are related to signals coupled into the waveguide. These phenomena are depicted on the bottom left of Figure 4.7. When the wave reaches the first slot, some

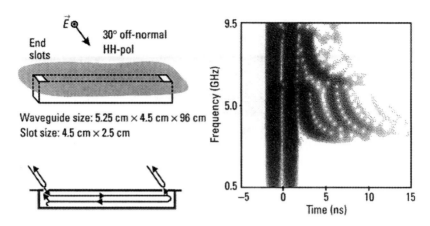

Figure 4.7 Joint time-frequency image of the back-scattering data from a slotted waveguide structure. The data were simulated using a method of moments solver.
The joint time-frequency image is generated by the STFT. The joint time-
frequency image shows both the early-time discrete-time returns from the slot
exterior and the late-time dispersive mechanisms due to modal propagation
inside the waveguide. (*Source:* [17] © 1995 IEEE.)

energy is coupled into the waveguide, propagating to the other end as a sum
of waveguide modes. The energy carried by these modes begins to reradiate
through the other slot after a time given by L/c, where L is the length
between the two slots. However, this is the time delay only for frequencies
well above the modal cutoff, for which the modal group velocity approaches
c. For frequencies approaching the cutoff frequency of the respective mode,
the group velocity tends to zero and the time delay goes to infinity. Consequently, each modal dispersion behavior is manifested as a time-frequency
trajectory with negative slope, as illustrated earlier in Figure 4.3(d). This
behavior can be clearly identified in the spectrogram, where the presence of
two modal dispersion curves with cutoffs at 3 GHz and at 6 GHz are
observed. They correspond to the TE_{10} and TE_{20} mode in the waveguide.
Note that the amplitude variation of the signal along these curves is governed
by the coupling mechanisms through the slot apertures and is considerably
more complex. Since multiple reflections occur, we also see other dispersion
curves with greater delays during the late-time portions of the return. The
first one corresponds with the energy that, upon reaching the other end,
reflects back and radiates through the slot on the left. The next is the three-
bounce mechanism. Note that energy is also coupled into the waveguide
through the slot on the right, and through similar mechanisms generate
dispersion curves in the time-frequency plot depending on the number of
bounces.

From the above four examples, we can see that the joint time-frequency representation can aid in the interpretation of complex electromagnetic scattering phenomena. Furthermore, the joint time-frequency features can be well understood in terms of the target scattering physics. For radar applications, the time-frequency representation is particularly effective for identifying scattering mechanisms in targets containing subskinline structures such as inlet ducts, antenna windows, and material coatings.

4.3 Application of High-Resolution Time-Frequency Techniques to Scattering Data

The additional insights gained in the time-frequency plane come at the price of resolution. As discussed in Chapter 2, the spectrogram generated by the STFT is limited in resolution by the extent of the sliding window function. Smaller time window results in better time resolution, but leads to worse frequency resolution, and vice versa. To overcome the resolution limit of the STFT, a wealth of alternative time-frequency representations have been developed by researchers in the signal processing community. Some of them were introduced in Chapter 2. In this section, we discuss how these techniques can be applied to analyze radar range profiles with improved feature resolution when compared to the STFT.

4.3.1 Use of the CWT

Contrary to the fixed resolution of the STFT, the wavelet transform is a time-frequency representation capable of achieving variable resolution in one domain (either time or frequency) and multiresolution in the other domain. We define here the CWT of a frequency signal $S(\omega)$ as

$$CWT(t, \omega) = \frac{1}{2\pi} \int S(\omega')t^{1/2} H(t(\omega' - \omega))d\omega' \qquad (4.5)$$

where $H(\cdot)$ is the mother wavelet and the resulting 2D magnitude display of (4.5) is the scalogram. The wavelet transform can also be carried out on the inverse Fourier transform $s(t)$ of the frequency signal $S(\omega)$

$$CWT(t, \omega) = \int s(t')t^{-1/2} h(-t'/t) \exp\{-j\omega t'\}dt' \qquad (4.6)$$

where $h(t')$ is the Fourier transform of $H(\omega')$. Since (4.6) is essentially the Fourier transform of $s(t')t^{-1/2}h(-t'/t)$, it is the preferred numerical implementation of the wavelet transform through the use of the fast Fourier transform (FFT) algorithm for each value of t. It is worthwhile to point out here that the definition of the wavelet transform presented above in its time and frequency forms is exactly the complement of its usual definition introduced in Section 2.1.2. For electromagnetic scattering applications, the property of the wavelet transform we are usually interested in is its multiresolution capability in the frequency domain and its variable resolution capability in the time domain. The multiresolution capability is ideally suited for analyzing frequency-domain electromagnetic back-scattering data that consist of both discrete time events (of large extent in frequency) and discrete frequency events (of small extent in frequency).

As an example, the time-frequency representation of the back-scattering data from an open-ended waveguide duct is considered [11]. The duct is an open-ended circular waveguide with a diameter of 1.75 in. A flat conducting termination exists 2 ft inside the waveguide [Figure 4.8(a)]. To generate the back-scattering data, the radar cross section of this target was first simulated in the frequency domain and the corresponding time-domain response was then obtained by inverse-Fourier transforming the band-limited frequency data (2–18 GHz). The polarization considered is the case where the magnetic field is polarized horizontally. Figure 4.8(b) shows the spectrogram of the back-scattering data at 45 degrees off-normal incidence using the STFT. Also plotted along the two axes are the time-domain and the frequency-domain responses. It is apparent that the scattering features are more revealing in the time-frequency domain than in either the time or the frequency domain alone. Both the nondispersive rim diffraction (i.e., the first vertical line) and the mode spectra due to multiple propagating modes in the circular waveguide can be clearly identified. However, due to the fixed resolution of the STFT, the scattering features are smeared out in the time-frequency plane. This problem is overcome by using the wavelet transform, as shown by the scalogram in Figure 4.8(c). The wavelet transform is implemented using (4.6) with the aid of the FFT. The function $h(t')$ is chosen to be a two-sided Kaiser-Bessel window with a Q-factor of 0.3. The $t = 0$ reference of $h(t')$ is located midway between the time events from the rim diffraction and interior contribution (at $t = 2.05$ nsec). The variable time resolution of the wavelet transform allows sharper time resolution to be achieved during the early-time response and sharper frequency resolution (coarser time resolution) to be achieved during the late-time response. Thus, wavelet transform provides good resolution in identifying the scattering centers and resolving the

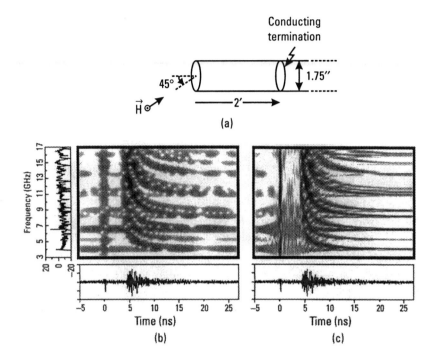

Figure 4.8 Joint time-frequency images of the back-scattered data from an open-ended waveguide duct obtained via the STFT and the CWT. (*Source:* [11] © 1993 IEEE.)

resonant phenomena of the target while adequately describing the dispersive scattering mechanisms in the intermediate-time region.

4.3.2 Use of the TFDS

While the STFT and the wavelet transform are based on linear transformations, another class of time-frequency distributions can be obtained from the quadratic, or power spectrum, point of view. The most basic of these is the WVD discussed in Section 2.2.1. Although the WVD gives nearly the best resolution among all the time-frequency techniques, its main drawback comes from cross-term interference problem. This drawback severely limits the usefulness of the WVD in its original form. A number of techniques have been proposed to alleviate this problem. Here we apply the TFDS proposed by Qian and Chen [18] and discussed in Section 2.2.3. In this technique, the original signal is first decomposed into its Gabor expansion. Then, by taking the WVD of the resulting expansion and selecting only the low-order harmonics, it is largely possible to preserve the resolution of the WVD while suppressing cross-term interference.

As an example, we consider the scattering from a dielectric coated wire [19]. The coated wire is a 12-inch section of a coaxial cable (RG-41/U with $a = 0.037$ in and $b = 0.095$ in) with its outer conductor removed. The dielectric material is Teflon with permittivity $\epsilon_r = 2.1$. The scattering data were collected by both numerical simulation and by chamber measurement. The spectrograms for the case of 60-degree incidence from broadside are shown in Figure 4.9(a). We observe a slight tilting of the vertical lines in the spectrograms, especially in the late-time returns. This signifies the presence of dispersive phenomena, which are due to the surface wave mechanism in the dielectric coating known as the Goubau mode. At high frequencies, this mode is tightly bound to the dielectric and the group velocity of the wave approaches the slow dielectric velocity. As the frequency approaches zero, the wave velocity approaches that of free space and exhibits a shorter propagation delay. Therefore, in the time-frequency plane, Goubau-mode phenomena show up as slanted curves with a positive slope. The TFDS and the WVD results are also given in Figure 4.9(b, c), respectively. As we can see

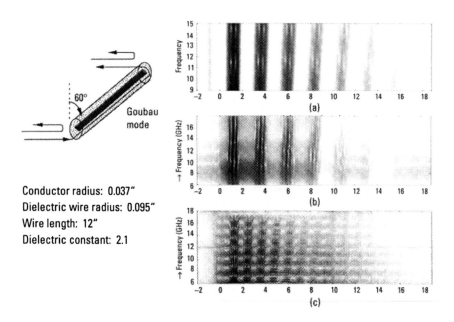

Figure 4.9 Joint time-frequency representations of the back-scattered data from a dielectric coated wire obtained via the STFT, the TFDS with $N = 2$, and the WVD. (*Source:* [19] © 1997 IEEE.)

from Figure 4.9(c), the WVD of the data is so contaminated by cross-term interference it is essentially useless. The TFDS result shown in Figure 4.9(b) is a good compromise between resolution and cross-term interference. The returns that could not be distinguished from each other in the spectrogram are now easily identified in the joint time-frequency plane by using the TFDS of order 2.

4.3.3 Windowed Superresolution Algorithm

In either the STFT or the wavelet transform, the resolution in the time-frequency plane is limited by the extent of the sliding window function. It is possible to achieve higher resolution by replacing the Fourier transform engine within the sliding window by a superresolution algorithm such as MUSIC [20] or ESPRIT [21] to process the data. Super-resolved parameterization retains the advantage of simultaneous time-frequency display while overcoming the resolution issue. However, additional processing is needed to fully parameterize the data, especially when dispersive mechanisms are present. Furthermore, the robustness of the algorithms to noise needs to be carefully considered. We describe here a simple windowed superresolution procedure based on Prony's method for achieving parameter estimation of both scattering centers and natural resonances in the time-frequency plane [14].

In the windowed time-frequency superresolution procedure, Prony's extraction is first applied in the frequency domain to locate discrete time events. Prony's method will fit the raw data to the following model

$$S(\omega) = \sum_{m=1}^{M} A_m \exp\{-j\omega t_m\} \qquad (4.7)$$

where the t_m's are the unknown locations of the discrete time events, the A_m's are the unknown complex strengths of these events, and M is the number of them to be found. It is clear that applying Prony's method to the entire frequency data will yield a poor fit if the raw data contains discrete frequency events such as resonances. To circumvent this problem, we repeatedly applied Prony's method to many small windows of the raw data. Prony's method will be successful for most of the window locations and will fail only when a window coincides with a resonant peak. By repeatedly sliding the window along in frequency and reapplying Prony's method to the raw data, we can identify as true locations those values of t_m that most frequently occur. A weighted-least squares fit of the values of A_m for each discrete time event is used to construct a smooth functional form of A_m. An important benefit of this global functional form, $A_m(\omega)$, is that it allows us to go back and interpolate A_m at those frequencies where Prony's method originally failed.

Provided that the discrete time events have been properly located and accounted for, the remaining data will consist solely of a series of resonant peaks after the frequency domain extraction. To extract the natural resonance information, this remaining data is first inverse-Fourier transformed to the time domain. The sliding window Prony's procedure is then applied in the time domain to fit the complex-valued data to a model that is the dual of (4.7)

$$s(t) = \sum_{n=1}^{N} B_n \exp\{j\omega_n t\} \qquad (4.8)$$

in which the ω_n's are the unknown resonance frequencies, and the B_n's are their corresponding strengths. By tracking the behavior of each B_n with respect to time, other parameters associated with the resonance, such as attenuation factor (α_n) and turn-on time (τ_n) are extracted

$$s(t) = \sum_{n=1}^{N} b_n \exp\{j\omega_n(t - \tau_n)\} \exp\{-\alpha_n(t - \tau_n)\} u(t - \tau_n) \qquad (4.9)$$

in which the b_n's are strengths of the resonances at turn-on and $u(\cdot)$ is the unit step function [i.e., $u(t) = 1$, for $t \geq 0$; and $u(t) = 0$, for $t < 0$].

As an example, let us again consider the perfectly conducting strip containing a small open cavity, shown earlier in Figure 4.4(a). From the spectrogram obtained via the STFT shown in Figure 4.4(d), it is possible to extract qualitative information such as the approximate frequency behaviors of the individual scattering centers and the resonance Q's. However, due to the large number of features contained in the back-scattered data, the image is blurry, making it difficult to resolve fine details. Figure 4.10 shows the time-frequency plot resulting from applying the windowed superresolution procedure discussed above to the back-scattered data. Because the data is fully parameterized via the superresolution approach, the sharpness of the image is not constrained by the well-known Fourier limit as is the case for the spectrogram. In the absence of noise, the image can be of nearly infinite sharpness, and each mechanism has been chosen to appear as either a horizontal or vertical line exactly one pixel in width. The intensities of the three

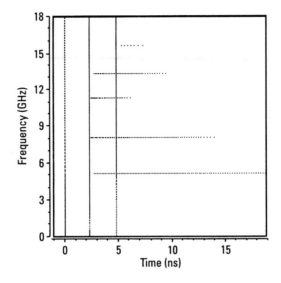

Figure 4.10 Time-frequency representation of the back-scattered data from a conducting strip with an open cavity obtained via a windowed superresolution algorithm. (*Source:* [14] © 1995 IEEE.)

vertical lines show that the three scattering centers are of differing strengths and have different frequency behaviors. The high-Q resonances can be seen to be of much longer duration than the low-Q ones. More robust superresolution algorithm may be used in place of Prony's method to achieve improved performance in the presence of noise [22]. The parameterization strategy can also be extended to deal with dispersive mechanisms [23, 24].

4.3.4 Adaptive Gaussian Representation

The wavelet transform is rather rigid in its particular form of the time-frequency grid. It is oftentimes desirable to achieve flexible resolution in the time-frequency plane to accommodate components of the signal with different resolutions. Several signal-adaptive time-frequency representations have been proposed in the literature, the best known of which are the adaptive Gaussian representation [25] and the matching pursuit algorithm [26]. The adaptive spectrogram, which was discussed in Section 2.1.3, uses adaptive normalized Gaussian functions to represent the signal. In the algorithm, the time and frequency resolution as well as the time-frequency centers are adjusted to best match the signal. The parameterization is carried out iteratively by projecting the signal onto all possible bases and selecting the one with the maximum projection value. That signal component is then subtracted from the original signal and the process is iterated until the original signal is adequately parameterized. The result of applying the adaptive Gaussian extraction can be effectively displayed in the time-frequency plane using the adaptive spectrogram. It is generated by calculating the WVD of the parameterized signal and deleting the cross terms.

To show an example of the adaptive spectrogram, we again use the same strip-cavity structure in Figure 4.4(a). From the spectrogram shown earlier in Figure 4.4(d), we observe that it is possible to extract qualitative information such as the approximate frequency behaviors of the individual scattering centers and the resonance Q's. However, due to the large number of features contained in the back-scattered data, the image is blurry, making it difficult to resolve fine details. We apply the adaptive Gaussian representation described in Section 2.1.3 and obtain an approximation of the signal using 50 terms with a residual energy of 0.1%. The resulting adaptive spectrogram is shown in Figure 4.11. We can clearly see and locate with high resolution the three scattering centers and the resonances. As expected, the high-Q resonances appear as very thin horizontal lines, while the low-Q resonances appear as thicker lines. We can even observe that the second and third scattering centers, which correspond respectively to the cavity exterior and the right fin, each contain multiple scattering events in time due to their more complicated shapes.

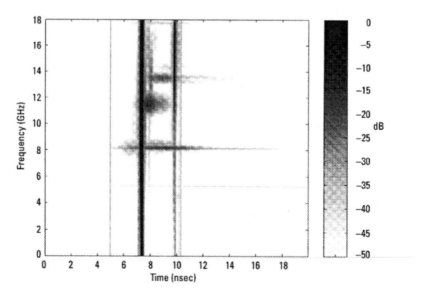

Figure 4.11 Adaptive spectrogram of the back-scattered data from the conducting strip with an open cavity shown in Figure 4.4(a).

4.4 Extraction of Dispersive Scattering Features from Radar Imagery Using Time-Frequency Processing

The joint time-frequency processing of 1D range profiles described in Sections 4.2 and 4.3 can be extended to deal with 2D radar imagery. ISAR imaging has long been used by the microwave radar community for object-diagnostic and target-identification applications. ISAR is a simple and very robust process for mapping the position and magnitude of the point-scatterers on a target from multifrequency, multiaspect back-scattered data. However, for complex targets containing other scattering phenomena such as resonances and dispersive mechanisms, image artifacts are often encountered in the resulting ISAR image [27]. One important example is the scattering from the engine inlet/exhaust duct on aircraft. It is a dominant contributor to the overall scattering from the target, yet its waveguide-like structure and the associated frequency-dependent scattering mechanisms make it a non-point-scattering feature. When processed and displayed by the conventional ISAR algorithm, the inlet return results in an image feature which is not well-focused, is not related to the spatial location of the scatterer, and can often obscure other important point features on the target. Therefore, it would be useful to automatically remove these artifacts from the ISAR image, leading to a cleaned ISAR image containing only physically meaningful point-scatterers. Furthermore, the extracted inlet features can be better displayed in a more meaningful feature space to identify target resonances and cutoff phenomena.

Joint time-frequency processing can be applied to ISAR image processing to accomplish the above objective [28]. The conceptual idea behind the joint time-frequency ISAR algorithm is to apply joint time-frequency transform to the range (or time) axis of the conventional range and cross-range ISAR image to gain an additional frequency dimension. The result is a three-dimensional (3D) range, cross-range, and frequency matrix, with each range and cross-range slice of this matrix representing an ISAR image at a particular frequency. This concept is illustrated in Figure 4.12. Consequently, by examining how the ISAR image varies with frequency, we can distinguish the frequency-independent scattering mechanisms from the frequency-dependent ones. In the actual implementation of the joint time-frequency ISAR, the choice of the joint time-frequency processing engine is critical to preserve range resolution. This is demonstrated below using the adaptive Gaussian representation discussed earlier in Sections 2.1.3 and 4.3.4.

The adaptive Gaussian representation has two distinct advantages over the STFT. First, it is a parametric procedure that results in very high time-

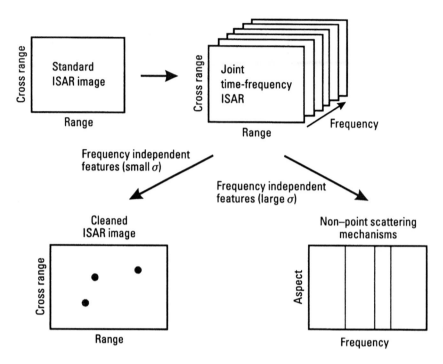

Figure 4.12 Joint time-frequency processing is applied to the range dimension of the conventional range and cross-range ISAR image to gain an additional frequency dimension. By examining how the resulting images vary as a function of frequency, the frequency-independent features can be separated from the frequency-dependent ones and displayed in an appropriate feature space. (*Source:* [29] © 1999 IEEE.)

frequency resolution. More importantly for the application under consideration, the adaptive representation allows us to automatically distinguish the frequency-dependent events from the frequency-independent ones through the extent of the basis functions. Equation (2.11) shows that scattering centers (i.e., signals with very narrow length in time) will be well represented by basis functions with very small σ_p. Frequency resonances, on the other hand, will be better depicted by large σ_p. Therefore, if we reconstruct the ISAR image using only those Gaussian bases with small variances, a much cleaner image can be obtained showing only the scattering centers. The remaining mechanisms (i.e., those related to the large-variance Gaussians) are more meaningful to view in a dual frequency-aspect display, where resonances and other frequency-dependent mechanisms can be better identified.

Two examples of joint time-frequency ISAR processing are presented. The first example is based on numerically simulated data for the perfectly conducting strip containing a small open cavity discussed earlier in Figure 4.4(a). Figure 4.13(a) shows its ISAR image at 30 degrees from edge on. The target outline is overlaid over the image for reference. The data used to form the image was collected from 5 to 15 GHz and over angular window of 0 to 60 degrees. We notice in the image that in addition to the three scattering centers corresponding to the left and right edge of the strip and the cavity exterior, there is a large cloud near the cavity spreading through the down range. This return corresponds to the energy coupled into the cavity and reradiated through the resonant mechanism. Figure 4.13(b) shows the enhanced ISAR image of Figure 4.13(a), obtained by applying the adaptive algorithm and keeping only the small-variance Gaussians. We see that the large cloud corresponding to the cavity resonance has been removed

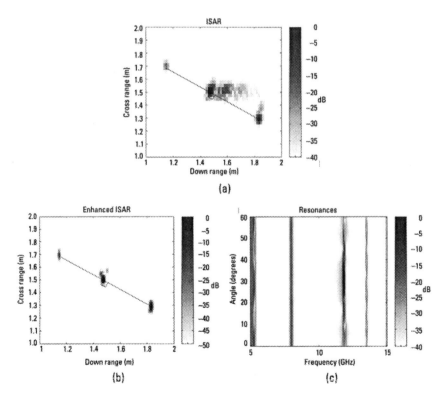

Figure 4.13 (a) The standard ISAR image of the conducting strip with a small cavity, obtained using frequency data from 5 to 15 GHz and angular data from 0 to 60 degrees; (b) enhanced ISAR image obtained by applying the adaptive Gaussian representation to the ISAR image in (a); and (c) the extracted resonant features of the inlet shown in the frequency-aspect plane. (*Source:* [28] © 1997 IEEE.)

and only the scattering center part of the original signal remains in the image, as expected. Figure 4.13(c) shows the frequency-aspect display of the high-variance Gaussians. Several very distinct equispaced vertical lines are observed. They correspond to the resonant frequencies of the cavity, which should occur at 5.30, 8.39, 11.86, and 13.52 GHz based on the dimensions of the cavity. Indeed, we see that they occur close to these frequencies and are almost aspect independent.

The algorithm is next demonstrated using the chamber measurement data of a 1:30 scale model Lockheed VFY-218 airplane provided by the Electromagnetic Code Consortium [30]. The airplane, shown in Figure 4.14(a), has two long engine inlet ducts that are rectangular at the open

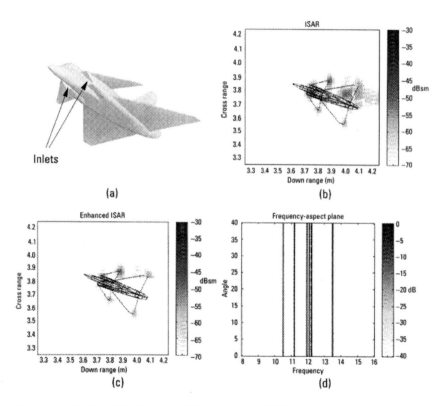

Figure 4.14 (a) The VFY-218 model; (b) its standard ISAR image obtained for f = 8 to 16 GHz and a 40-degree angular window centered at 30 degrees from nose-on; (c) enhanced ISAR image obtained by applying the adaptive Gaussian representation to the ISAR image (the inlet cloud has been removed from the original ISAR image); and (d) the extracted resonant features of the inlet are shown in the frequency-aspect plane. (*Source:* [31] © 1997 IEEE.)

ends but merge together into one circular section before reaching a single-compressor face. As we can clearly see in the conventional ISAR image of Figure 4.14(b) for the horizontal polarization at 20 degrees near nose-on, the large cloud outside of the airframe structure is the inlet return. Figure 4.14(c) shows the enhanced ISAR image of Figure 4.14(b), obtained by applying the joint time-frequency ISAR algorithm and keeping only the small-variance Gaussians. We see that only the scattering center part of the original signal remains in the image, as expected. Notice that the strong return due to engine inlet has been removed, but the scattering from the right wing tip remains. Figure 4.14(d) shows the frequency-aspect display of the high-variance Gaussians. A number of equispaced vertical lines can be seen between 10.5 and 13.5 GHz. Given the dimension of the rectangular inlet opening, we estimate that these frequencies correspond approximately to the second cutoff frequency of the waveguide-like inlet. This information is unique to the particular inlet structure under consideration and may be useful as an additional feature vector in target classification [32].

In summary, we have discussed a joint time-frequency ISAR algorithm to process data from complex targets containing not only scattering centers but also other frequency-dependent scattering mechanisms. The adaptive joint time-frequency ISAR algorithm allows the enhancement of the ISAR image by eliminating non-point-scatterer signals, thus leading to a much cleaner ISAR image. It also provides information on the extracted frequency-dependent mechanisms such as resonances and frequency dispersions. This is accomplished without any loss in resolution.

References

[1] Wehner, D. R., *High-Resolution Radar, Second Edition,* Norwood, MA: Artech House, 1994.

[2] Keller, J. B., "Geometrical Theory of Diffraction," *J. Opt. Soc. Am.,* Vol. 52, January 1962, pp. 116–130.

[3] Kouyoumjian, R. G., "The Geometrical Theory of Diffraction and Its Applications," *Numerical and Asymptotic Techniques in Electromagnetics,* R. Mittra (ed.), New York: Springer-Verlag, 1975.

[4] Kouyoumjian, R. G., and P. H. Pathak, "A Uniform Geometrical Theory of Diffraction for an Edge in a Perfectly Conducting Surface," *Proc. IEEE,* Vol. 62, November 1974, pp. 1448–1461.

[5] Lee, S. W., and G. A. Deschamps, "A Uniform Asymptotic Theory of EM Diffraction by a Curved Wedge," *IEEE Trans. Antennas Propagat.,* Vol. AP-24, January 1976, pp. 25–34.

[6]   Potter, L. C., et al., "A GTD-Based Parametric Model for Radar Scattering," *IEEE Trans. Antennas Propagat.*, Vol. 43, October 1995, pp. 1058–1067.

[7]   Cohen, L., *Time-Frequency Analysis*, Englewood Cliffs, NJ: Prentice Hall, 1995.

[8]   Qian, S., and D. Chen, *Introduction to Joint Time-Frequency Analysis—Methods and Applications*, Englewood Cliffs, NJ: Prentice Hall, 1996.

[9]   Casey, K. F., D. G. Dudley, and M. R. Portnoff, "Radiation and Dispersion Effects from Frequency-Modulated (FM) Sources," *Electromagnetics*, Vol. 10, 1990, pp. 349–376.

[10] Moghaddar, A., and E. K. Walton, "Time-Frequency-Distribution Analysis of Scattering from Waveguide Cavities," *IEEE Trans. Antennas Propagat.*, Vol. AP-41, May 1993, pp. 677–680.

[11] Kim, H., and H. Ling, "Wavelet Analysis of Radar Echo from Finite-Size Targets," *IEEE Trans. Antennas Propagat.*, Vol. AP-41, February 1993, pp. 200–207.

[12] Carin, L., et al., "Wave-Oriented Signal Processing of Dispersive Time-Domain Scattering Data," *IEEE Trans. Antennas Propagat.*, Vol. AP-45, April 1997, pp. 592–600.

[13] Ling, H., "Joint Time-Frequency Processing of Electromagnetic Backscattered Data," *SPIE Proc. on Wavelet Applications*, Orlando, FL, Vol. 3391, April 1998, pp. 283–294.

[14] Moore, J., and H. Ling, "Super-Resolved Time-Frequency Analysis of Wideband Backscattered Data," *IEEE Trans. Antennas Propagat.*, Vol. AP-43, June 1995, pp. 623–626.

[15] Ling, H., et al., "Time-Frequency Analysis of Backscattered Data from a Coated Strip with a Gap," *IEEE Trans. Antennas Propagat.*, Vol. AP-41, August 1993, pp. 1147–1150.

[16] Moore, J., and H. Ling, "Time-Frequency Analysis of the Scattering Phenomenology in Finite Dielectric Gratings," *Microwave Optical Tech. Lett.*, Vol. 6, August 1993, pp. 597–600.

[17] Trintinalia, L. C., and H. Ling, "Interpretation of Scattering Phenomenology in Slotted Waveguide Structures via Time-Frequency Processing," *IEEE Trans. Antennas Propagat.*, Vol. AP-43, November 1995, pp. 1253–1261.

[18] Qian, S., and D. Chen, "Decomposition of the Wigner-Ville Distribution and Time-Frequency Distribution Series," *IEEE Trans. Signal Processing*, Vol. 42, October 1994, pp. 2836–2842.

[19] Ozdemir, C., and H. Ling, "Interpretation of Scattering Phenomenology in Dielectric-Coated Wires via Joint Time-Frequency Processing," *IEEE Trans. Antennas Propagat.*, Vol. AP-45, August 1997, pp. 1259–1264.

[20] Schmidt, R. O., "Multiple Emitter Location and Signal Parameter Estimation," *IEEE Trans. Antennas Propagat.*, Vol. AP-34, March 1986, pp. 276–280.

[21] Roy, R., A. Paulraj, and T. Kailath, "ESPRIT—A Subspace Rotation Approach to Estimation of Parameters of Cisoids in Noise," *IEEE Trans. Acoust., Speech, Signal Processing*, Vol. ASSP-34, October 1986, pp. 1340–1342.

[22] Moore, J., et al., "Super-Resolved Time-Frequency Processing of Wideband Radar Echo Using ESPRIT," *Microwave Optical Tech. Lett.*, Vol. 9, May 1995, pp. 17–19.

[23] Moore, J., and H. Ling, "Super-Resolved Time-Frequency Processing of Surface Wave Mechanisms Contained in Wideband Radar Echo," *Microwave Optical Tech. Lett.*, Vol. 9, August 1995, pp. 237–240.

[24] Trintinalia, L. C., and H. Ling, "Super-Resolved Time-Frequency Parameterization of Electromagnetic Scattering Mechanisms Due to Structural Dispersion," *Microwave Optical Tech. Lett.*, Vol. 10, October 1995, pp. 82–84.

[25] Qian, S., and D. Chen, "Signal Representation Using Adaptive Normalized Gaussian Functions," *Signal Processing*, Vol. 36, No. 1, March 1994, pp. 1–11.

[26] Mallat, S. G., and Z. Zhang, "Matching Pursuits with Time-Frequency Dictionaries," *IEEE Trans. Signal Processing*, Vol. 41, December 1993, pp. 3397–3415.

[27] Borden, B., *Radar Imaging of Airborne Targets: A Primer for Applied Mathematicians and Physicists*, London, England: Inst. of Physics Pub., 1999.

[28] Trintinalia, L. C., and H. Ling, "Joint Time-Frequency ISAR Using Adaptive Processing," *IEEE Trans. Antennas Propagat.*, Vol. AP-45, February 1997, pp. 221–227.

[29] Chen, V. C., and H. Ling, "Joint Time-Frequency Analysis for Radar Signal and Image Processing," *IEEE Signal Processing Mag.*, Special Issue on Joint Time-Frequency Analysis, Vol. 16, March 1999, pp. 81–93.

[30] Wang, H. T. G., M. L. Sanders, and A. Woo, "Radar Cross Section Measurement Data of the VFY 218 Configuration," *Tech. Rept. NAWCWPNS TM-7621*, China Lake, CA: Naval Air Warfare Center, January 1994.

[31] Ling, H., Y. Wang, and V. Chen, "ISAR Image Formation and Feature Extraction Using Adaptive Joint Time-Frequency Processing," *SPIE Proc. on Wavelet Applications*, Orlando, FL, Vol. 3708, April 1997, pp. 424–432.

[32] Borden, B., "An Observation About Radar Imaging of Re-Entrant Structures with Implications for Automatic Target Recognition," *Inverse Problems*, Vol. 13, 1997, pp. 1441–1452.

5

Time-Frequency-Based Radar Image Formation

Radar image formation is a process of reconstructing images of radar targets from recorded complex data. All imaging techniques, essentially, project a 3D object space onto a 2D image plane. Radar image, specifically, is a mapping of a 3D target onto a 2D range and cross-range plane. To generate an image with radar systems, three major components (transmitter, target, and receiver) are required. The transmitter emits a sequence of pulses to the target to be imaged; the receiver then records the reflected pulses from the target and processes the recorded data to reconstruct an image of the target. To generate a 2D radar range and cross-range image, the recorded raw data need to be rearranged into a 2D format. The range resolution of a radar image is directly related to the bandwidth of the transmitted radar signal, and the cross-range resolution is determined by the effective antenna beamwidth, which is inversely proportional to the effective length of the antenna aperture. As we introduced in Chapter 1, to achieve a high cross-range resolution without using a large physical antenna aperture, synthetic array processing is widely used which coherently combines signals obtained from sequences of small apertures at different angle aspects to a target to emulate the result from a large aperture.

As we mentioned in Chapter 1, SAR generates a high-resolution map of stationary surface targets and terrain; ISAR uses a geometrically inverse way where the radar is stationary and targets are moving to generate image of targets [1–3]. With a sufficient high-Doppler resolution, differential

Doppler shifts of adjacent scatterers on a target can be observed, and the distribution of the target's reflectivity can be obtained through the Doppler frequency spectrum. Conventional methods to retrieve Doppler information are based on the Fourier transform. By taking the Fourier transform of a sequence of time history series, an ISAR image can be formed. Therefore, the conventional radar image formation is a Fourier-based image formation [4].

In this chapter, we discuss the Fourier-based image formation and introduce a new time-frequency-based image formation. We will briefly describe the background of radar imaging of moving targets and the time-varying behavior of Doppler frequency shifts in Section 5.1. We will discuss the standard motion compensation and image formation in Section 5.2 and introduce the time-frequency-based image formation in Section 5.3. Some issues on radar imaging of maneuvering targets and multiple targets will be discussed in Sections 5.4 and 5.5.

5.1 Radar Imaging of Moving Targets

The geometry of the radar imaging of a target is shown in Figure 5.1. The radar is located at the origin of the Cartesian coordinates (U, V, W), called the radar coordinates. The target is described in Cartesian coordinates (x, y, z) with its origin located at the geometric center of the target, called target coordinates. To describe rotations of the target, new reference coordinates (X, Y, Z), translated from the radar coordinates (U, V, W) and with origin at the geometric center of the target, are introduced. For simplicity, we show only a planar target in 2D coordinates. The third dimension can be easily added with the necessary equations modified by an elevation angle.

Assume that the radar transmits a sinusoidal waveform with a carrier frequency f_0. At time $t = 0$, the target range (i.e., the distance from the radar antenna to the geometric center of the target) is R, and the distance from the radar to a point-scatterer P on the target, located at $(x, y, z = 0)$, is

$$
\begin{aligned}
R_P &= [(T_X + x \cos\theta_0 - y \sin\theta_0)^2 + (T_Y + y \cos\theta_0 + x \sin\theta_0)^2]^{1/2} \\
&= \{R^2 + (x^2 + y^2) + 2R[x \cos(\theta_0 - \alpha) - y \sin(\theta_0 - \alpha)]\}^{1/2} \quad (5.1) \\
&\cong R + x \cos(\theta_0 - \alpha) - y \sin(\theta_0 - \alpha)
\end{aligned}
$$

where $(T_X, T_Y, T_Z = 0)$ is the translation of the origin of the (x, y, z) coordinates with respect to the radar (U, V, W) coordinates, α is the azimuth

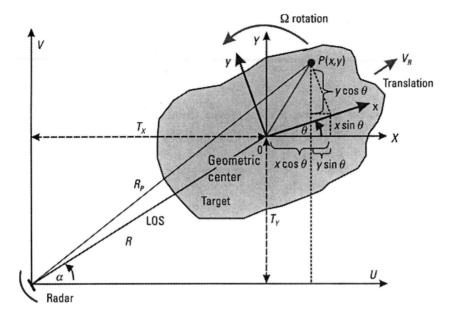

Figure 5.1 Geometry of the radar imaging of a target.

angle of the target with respect to the (U, V, W) coordinates, and θ_0 is the initial rotation angle of the (x, y, z) coordinates about the Z-axis in the (X, Y, Z) coordinates as illustrated in Figure 5.1.

If the target has a rotational motion with an initial angular rotation rate Ω about the Z-axis and a translational motion with a radial velocity V_R, then the range and the rotation angle of the target is a function of time. The range can be expressed by the target's initial range R_0, the initial velocity V_R, the initial radial acceleration $a_R(= dV_R/dt)$, and other higher order terms; and the rotation angle can be expressed by the initial orientation angle θ_0 with respect to the (X, Y, Z) coordinates, the initial angular rotation rate Ω, the initial angular acceleration $\gamma(= d\Omega/dt)$, and other higher order terms:

$$R(t) = R_0 + V_R t + \frac{1}{2} a_R t^2 + \ldots \tag{5.2}$$

and

$$\theta(t) = \theta_0 + \Omega t + \frac{1}{2} \gamma t^2 + \ldots \tag{5.3}$$

Thus, the range from the radar to the point-scatterer P becomes

$$R_P(t) = R(t) + x \cos[\theta(t) - \alpha] - y \sin[\theta(t) - \alpha] \qquad (5.4)$$

and the baseband of the returned signal from the point-scatterer P is a function of $R_t = R(t)$ and $\theta_t = \theta(t)$:

$$s_P(t) = \rho(x, y, z) \exp\left\{ j2\pi f_0 \frac{2R_P(t)}{c} \right\} = \rho(x, y, z) \exp\{j\Phi(R_{P_t})\}$$

$$(5.5)$$

where R_{P_t} is an abbreviation of $R_P(t)$, $\rho(x, y, z)$ is the reflectivity function of the point-scatterer P at (x, y, z), and c is the speed of electromagnetic wave propagation.

The phase of the baseband signal is

$$\Phi(R_{P_t}) = 2\pi f_0 \frac{2R_P(t)}{c} \qquad (5.6)$$

By taking the time-derivative of the phase, the Doppler frequency shift induced by the target's motion is approximately

$$f_D = \frac{2f_0}{c} \frac{d}{dt} R_P(t)$$

$$= \frac{2f_0}{c} V_R + \frac{2f_0}{c} [-x\Omega \sin(\theta_0 + \Omega t - \alpha) - y\Omega \cos(\theta_0 + \Omega t - \alpha)]$$

$$= \frac{2f_0}{c} V_R + \frac{2f_0}{c} \{-x\Omega[\sin(\theta_0 - \alpha) \cos\Omega t + \cos(\theta_0 - \alpha) \sin\Omega t]$$

$$- y\Omega[\cos(\theta_0 - \alpha) \cos\Omega t - \sin(\theta_0 - \alpha) \sin\Omega t]\} \qquad (5.7)$$

where we only use the zero and first-order terms in (5.2) and (5.3). For a given rotation rate and coherent processing time t, if $\Omega^2 t^2 \ll 1$ and $\Omega^3 t^3 \ll \Omega t$, hence $\cos\Omega t = 1 - \Omega^2 t^2/2 + \ldots \cong 1$ and $\sin\Omega t \cong \Omega t - \Omega^3 t^3/6 + \ldots \cong \Omega t$, we have

$$f_D \cong \frac{2f_0}{c} V_R + \frac{2f_0}{c} \{-x\Omega[\sin(\theta_0 - \alpha) + \cos(\theta_0 - \alpha)\Omega t]$$

$$-y\Omega[\cos(\theta_0 - \alpha) - \sin(\theta_0 - \alpha)\Omega t]\}$$

$$= \frac{2f_0}{c} V_R + \frac{2f_0}{c} \{-[x\sin(\theta_0 - \alpha) + y\cos(\theta_0 - \alpha)]\Omega$$

$$- [x\cos(\theta_0 - \alpha) - y\sin(\theta_0 - \alpha)]\Omega^2 t\} = f_{D_{Trans}} + f_{D_{Rot}} \quad (5.8)$$

where the Doppler frequency shift induced by the translational motion is

$$f_{D_{Trans}} = \frac{2f_0}{c} V_R \quad (5.9)$$

and that induced by the rotational motion is

$$f_{D_{Rot}} = \frac{2f_0}{c} \{-[x\sin(\theta_0 - \alpha) + y\cos(\theta_0 - \alpha)]\Omega \quad (5.10)$$

$$- [x\cos(\theta_0 - \alpha) - y\sin(\theta_0 - \alpha)]\Omega^2 t\}$$

The first and the second terms of (5.10) come from the linear and quadratic parts of the phase function, respectively. The quadratic part of the rotational Doppler frequency shift is a function of time. Therefore, given angular rotation rate, carrier frequency, and the scatterer's location (x, y, z), if $\frac{2f_0}{c} [x\cos(\theta_0 - \alpha) - y\sin(\theta_0 - \alpha)]\Omega^2$ cannot be neglected, the rotational Doppler frequency shift is time-varying, even if the angular rotation rate Ω is a constant.

Based on the returned signal from a single point-scatterer, the returned signal from the target can be represented as the integration of the returned signals from all scatterers in the target:

$$s_R(t) = \int_{-\infty}^{\infty} \int_{-\infty}^{\infty} \int_{-\infty}^{\infty} \rho(x, y, z) \exp\left\{-j2\pi f_0 \frac{2R_P(t)}{c}\right\} dx dy dz \quad (5.11)$$

$$for\ 2R_P(t)/c \le t \le T_{PRI} + 2R_P(t)/c$$

where T_{PRI} is the PRI of the transmitted signal.

For a target that has translational and rotational motion and, for simplicity, assuming the target's azimuth angle α is zero, then the range of a point-scatterer at $(x, y, z = 0)$ in the target coordinate system can be rewritten as $R_P(t) = R(t) + x \cos \theta(t) - y \sin \theta(t)$ and the returned signal in (5.11) can be rewritten as

$$s_R(t) = \exp\{-j4\pi f_0 R(t)/c\} \int_{-\infty}^{\infty} \int_{-\infty}^{\infty} \int_{-\infty}^{\infty} \rho(x, y, z) \quad (5.12)$$

$$\exp\left\{-j2\pi[xf_x(t) - yf_y(t)]\right\} dx\,dy\,dz$$

$$\text{for } 2R_P(t)/c \leq t \leq T_{PRI} + 2R_P(t)/c$$

where the components of the spatial frequency are determined by

$$f_x(t) = \frac{2f_0}{c} \cos \theta(t) \quad (5.13)$$

and

$$f_y(t) = \frac{2f_0}{c} \sin \theta(t) \quad (5.14)$$

From (5.12) we know that if the target's initial range R_0 is known exactly and the velocity V_R and acceleration a_R of the target's motion are known exactly over the entire coherent processing interval, then the extraneous phase term of the motion $\exp\{-j4\pi f_0 R(t)/c\}$ can be exactly removed by multiplying $\exp\{j4\pi f_0 R(t)/c\}$ on both sides of (5.12). Therefore, the reflectivity density function $\rho(x, y, z)$ of the target can be obtained simply by taking the inverse Fourier transform of the phase-compensated baseband signal $s_R(t) \exp\{j4\pi f_0 R(t)/c\}$.

The process of estimating the target's motion and removing the extraneous phase term is called range tracking. This is a fundamental step in the standard motion compensation procedure, also called coarse motion compensation. Then, the inverse Fourier transform may be used to reconstruct the reflective density function of the target.

In order to use the Fourier transform properly, certain conditions must be satisfied. During the entire coherent imaging processing time, the scatterers must remain in their range cells, and their Doppler frequency shifts must be constant. If the scatterers drift out their range cells or their Doppler frequency shifts are time-varying, the image reconstructed by using the Fourier transform becomes blurred. Only with the range tracking processing and without applying any phase compensation, can Doppler frequency shifts still be time-varying. Thus, a fine motion compensation, called Doppler tracking, should be applied to make phase compensation and, hence, Doppler frequency shifts, constant. The range tracking and Doppler tracking are the bases of the standard motion compensation.

Figure 5.2 illustrates the process of the ISAR imaging system using a wide-band waveform. The radar transmits a sequence of N pulses. The range resolution is determined by the bandwidth of the pulse. For each transmitted pulse, the total number of range cells, M, is determined by the maximum range covered and the range resolution. The total number of pulses, N, for a given imaging integration time determines the Doppler or cross-range resolution.

After being pulse-compressed, heterodyned, and quadraturely detected in the radar receiver, the base-band I and Q signals as defined in Chapter 1 are organized into an $M \times N$ 2D complex array $s_R(r_{m,n})$ where $m = 0$, $1, \ldots, M - 1$; $n = 0, 1, \ldots, N - 1$. Therefore N range profiles, each containing M range cells, can be obtained. At each range cell, the data across the N range profiles constitutes a new time history series. After applying range tracking and Doppler tracking, the aligned-range profiles become $G(r_{m,n})$, $(m = 0, 1, \ldots, M - 1; n = 0, 1, \ldots, N - 1)$.

The Fourier-based image formation takes the Fourier transform or FFT for the new time history series and generates an N-point Doppler spectrum called the Doppler profile. By combining the M Doppler spectra at M range cells, finally, the $M \times N$ image is formed

$$I(r_m, f_n) = FFT_n\{G(r_{m,n})\} \tag{5.15}$$

where FFT_n denotes the FFT operation with respect to the variable n. Therefore the radar image $I(r_m, f_n)$ is the target's reflectivities mapped onto the range-Doppler plane.

As we described earlier in (5.10), when the target's angular rotation rate is high or the image coherent processing interval is long, the rotational Doppler frequency shift can be time varying.

Other sources of time-variation in the Doppler frequency shift may result from uncompensated phase errors due to irregularities in the motion of the target, the fluctuation of the rotation rate of the target, fluctuation in localizing the rotation center, inaccuracy in tracking the phase history,

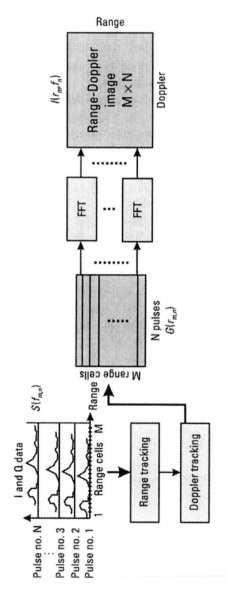

Figure 5.2 Illustration of the process of the SAR imaging system.

and other variations of the system and the environment. From the relationship between the range and the phase given in (5.6), the phase is very sensitive to the range variation. For example, for an X-band radar at 10 GHz, $\Delta r = 1.5$ cm (i.e., half-wavelength) range drift can cause $\Delta \Phi = 4\pi f_0 \Delta r/c = $ 360-degree phase deviation. Since the residual phase errors may vary with time, the Doppler frequency also varies with time.

As is known [5, 6], the Fourier transform only indicates what frequency components are contained in the signal; it does not tell how frequencies change with time. By representing the time-varying Doppler frequency spectrum with the Fourier transform, the Doppler spectrum becomes smeared. For showing this, we apply the Fourier transform and the STFT [7, 8] to a time history series of a measured radar data as shown in Figure 5.3. We can see that the Fourier transform of the time history series is actually the integral of the time-frequency transform of the same series over its time

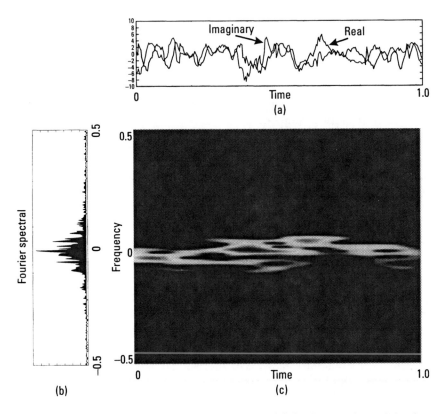

Figure 5.3 (a) Time history series of real radar data; (b) Fourier transform of the time history series; and (c) time-frequency transform of the same time history series.

duration. This is because of the frequency marginal condition. For a signal $s(t)$, if its joint time-frequency energy distribution $P(t, f)$ satisfies the following condition:

$$\int P(t, f)dt = |S(f)|^2 \qquad (5.16)$$

and

$$\int P(t, f)df = |s(t)|^2 \qquad (5.17)$$

where $S(f)$ is the Fourier transform of the signal, the time-frequency transform satisfies the frequency and the time marginal conditions.

Because the cause of image smearing is the use of the Fourier transform for such data that has time-varying Doppler frequency spectrum, to better deal with such data and generate a clear image, a time-frequency transform can be used to replace the Fourier transform. The time-frequency transform introduced in Chapter 2 is an efficient way to resolve the image smearing caused by the time-varying Doppler's behavior without applying sophisticated algorithms of motion compensation.

5.2 Standard Motion Compensation and Fourier-Based Image Formation

To generate a clear radar image, motion compensation algorithms must be applied. The purpose of the motion compensation is to preprocess the data such that conventional Fourier image formation can be applied to obtain a well-focused radar image.

As is described in Section 5.1, to use the Fourier-based image formation properly, the following conditions must be satisfied during the coherent image processing time: (1) the scatterers on the target must remain in their range cells, and (2) their Doppler frequency shifts must be constant. If the scatterers drift out their range cells or their Doppler frequency shifts are time varying, the Doppler spectrum obtained from the Fourier transform becomes smeared, and the radar image becomes blurred.

Standard motion compensation includes range tracking (by applying range-bin alignment) and Doppler tracking (by applying phase correction or compensation) [9–15]. Range tracking can keep scatterers in their range cells; Doppler tracking makes Doppler frequency shifts to be constant as illustrated in Figure 5.4. Therefore, after motion compensation, all scatterers on the target appear to be moving with a constant speed (due to its constant Doppler frequency shift) and along a perfect circle (due to its constant range). The range tracking or alignment process can be performed by a cross-correlation method that finds misaligned range cells with respect to a reference range profile and, then, performs range alignment for all range profiles. The Doppler tracking is performed using a phase compensation method. The phase compensation procedure usually includes three steps: (1) searching for one or several reference range cells by using a criteria such as minimum variance; (2) taking conjugate phase at the reference range cells; and (3) making phase correction for all range cells using the conjugate phase. Figure 5.4 illustrates the standard motion compensation diagram and shows results of the range tracking and Doppler tracking.

If a target is moving smoothly, standard motion compensation is good enough to generate a clear image of the target by using the Fourier transform. However, when a target exhibits complex motion such as rotation, acceleration or maneuvering, the standard motion compensation is not sufficient to generate an acceptable image for viewing and analysis. In that case, more sophisticated algorithms for compensating motions of individual scatterers, such as polar reformatting and other more complicated algorithms, are needed. Thus, each scatterer can remain in its range cell and its Doppler

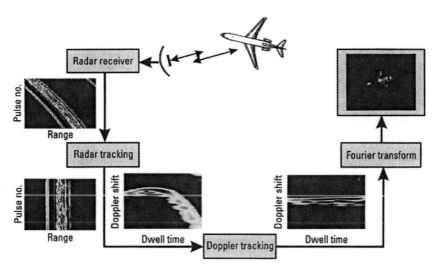

Figure 5.4 Standard motion compensation diagram.

frequency shift becomes a constant. Then the Fourier transform can be applied properly to reconstruct a clear image of the target [16].

Polar reformatting, which can correct rotational motion for individual scatterers, requires one to resample the data so that the sample points on the polar sampling grid are conformed to the desired sample points on a rectangular sampling grid [1, 2]. Besides, to perform the polar reformatting some initial kinematic parameters of the target are required.

If the sophisticated motion compensation is still not sufficient, these individual scatterers may still drift through their range cells, and their Doppler frequency shifts may still be time-varying. Thus, the resulting image can still be blurred if the conventional Fourier transform is applied.

However, the restriction of the Fourier transform can be circumvented if it is replaced with a time-frequency transform. Because of the time-varying behavior of the Doppler frequency shift, an efficient method to solve the problem of the smeared Fourier frequency spectrum and, hence, the blurred image is to apply a high-resolution time-frequency transform to the Doppler processing. In that way, the image blurring caused by time-varying Doppler frequency shifts can be mitigated without applying sophisticated motion compensation algorithms [17].

5.3 Time-Frequency-Based Image Formation

We described the basic concept of ISAR imaging in Chapter 1 and at the beginning of this chapter, and also described the conventional Fourier-based imaging system in Section 5.1. To generate a clear ISAR image of maneuvering targets, a time-frequency transform with superior resolution, low cross-term interference, and unbiased estimation of the instantaneous frequency spectrum is always desirable.

To apply the time-frequency-based image formation, we need a time-frequency transform that is specially designed for computing time-varying spectra and retrieving instantaneous Doppler frequency information. Having a high-resolution time-varying Doppler spectrum, it is no longer necessary to flatten out the distribution of the Doppler frequency spectrum and to compensate motions of individual scatterers for obtaining a clear image of moving targets. Instead of generating a range and Doppler (or cross-range with a known scaling factor) image, the time-varying Doppler spectrum can be used to generate a number of range and instantaneous Doppler images.

Figure 5.5 illustrates the radar imaging system based on the time-frequency transform [17–21]. The only difference between the time-

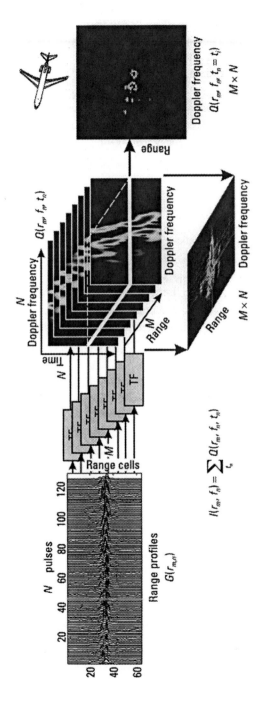

Figure 5.5 Illustration of a radar imaging system based on the time-frequency image formation.

frequency-based imaging system and the conventional Fourier-based imaging system is that the Fourier transform is replaced by a time-frequency transform, followed by time sampling. Assuming the data is formed as a complex 2D array $G(r_{m,n})$ with M time history series, each having the length of N (or N pulses), the Fourier-based imaging formation generates only one image frame from the $M \times N$ I and Q data array. However, the time-frequency-based imaging formation takes a time-frequency transform for each time history series and generates an $N \times N$ time-Doppler distribution. By combining the M time-Doppler distributions at M range cells, the $N \times M \times N$ time-range-Doppler cube $Q(r_m, f_n, t_n)$ can be formed:

$$Q(r_m, f_n, t_n) = TFT_n\{G(r_{m,n})\}$$

where TFT_n denotes the time-frequency transform with respect to the variable n.

Because the time-frequency transform can calculate the instantaneous Doppler frequency shift, at any instant the Doppler frequency shift of each scatterer on a target becomes a fixed value with its Doppler resolution determined by the selected time-frequency transform. At a sampling time t_i, only one range and instantaneous Doppler image frame $Q(r_m, f_n, t_n = t_i)$ can be extracted from the $N \times M \times N$ time-range-Doppler cube. There are a total of N image frames available, and each frame represents a full range-Doppler image at a particular time. Therefore, by replacing the Fourier transform with the time-frequency transform, a 2D range-Doppler image becomes a 3D time-range-Doppler image cube. By taking time sampling, a temporal sequence of 2D range-Doppler images can be viewed [17]. Each individual time-sampled frame in the cube provides a clear image with a superior resolution provided by the time-frequency transform. According to the frequency marginal condition in (5.16), integration of the N frames turns out to be a Fourier image:

$$I(r_m, f_n) = \sum_{t_n = t_0}^{t_{N-1}} Q(r_m, f_n, t_n) \qquad (5.18)$$

It is not usually necessary to take the maximal N time-samples because the Doppler variation from one sample to the next is not significant. In many cases, 16 or 32 equally spaced time-samples may be good enough to show the detailed Doppler variations.

In the following sections, we will give examples of the time-frequency transform for radar imaging of maneuvering targets and multiple targets.

5.4 Radar Imaging of Maneuvering Targets

A maneuvering target is defined as one that has translational and rotational, possibly nonuniform motions during the coherent processing interval. A challenge in radar imaging is how to form a clear image of maneuvering targets. In this section, we will describe dynamics of maneuvering targets and apply time-frequency-based image formation to radar imaging of maneuvering targets. We will use both the simulated and the measured radar data to demonstrate the time-frequency-based image formation. We also compare different time-frequency transforms used in the time-frequency-based image formation.

5.4.1 Dynamics of Maneuvering Targets

For a coordinate system as that shown earlier in Figure 5.1, where the target coordinates (x, y, z) have a translation (T_X, T_Y, T_Z) from the radar coordinates (U, V, W) and has a rotation angle $\theta(t)$ at time t about the reference coordinates (X, Y, Z), which have translation only from the (U, V, W) coordinates but no translation from the (x, y, z) coordinates. Thus, the translation matrix is defined as

$$Trans(T_X, T_Y, T_Z) = \begin{bmatrix} 1 & 0 & 0 & T_X \\ 0 & 1 & 0 & T_Y \\ 0 & 0 & 1 & T_Z \\ 0 & 0 & 0 & 1 \end{bmatrix} \quad (5.19)$$

where $T_X = R\cos\alpha$, $T_Y = R\sin\alpha$, and $T_Z = 0$ as shown in Figure 5.1.

A frequently used set of rotational motions is roll, pitch, and yaw. For an aircraft heading along the x-axis, roll corresponds to a rotation about the x-axis, pitch corresponds to a rotation about the y-axis, and yaw corresponds to a rotation about the z-axis. If the order of rotations is a roll with an angle θ_r, followed by a pitch with an angle θ_p, and finally, a yaw with an angle θ_y, then the composite roll, pitch, and yaw motion in the target's coordinates can be represented by a rotation matrix [21].

$$Rot(\theta_r, \theta_p, \theta_y) = \begin{bmatrix} a_{11} & a_{12} & a_{13} \\ a_{21} & a_{22} & a_{23} \\ a_{31} & a_{32} & a_{33} \end{bmatrix} \quad (5.20)$$

where

$$a_{11} = \cos\theta_p \cos\theta_y$$
$$a_{12} = -\cos\theta_p \sin\theta_y$$
$$a_{13} = \sin\theta_p$$
$$a_{21} = \sin\theta_r \sin\theta_p \cos\theta_y + \cos\theta_r \sin\theta_y$$
$$a_{22} = -\sin\theta_r \sin\theta_p \sin\theta_y + \cos\theta_r \cos\theta_y$$
$$a_{23} = -\sin\theta_r \cos\theta_p$$
$$a_{31} = -\cos\theta_r \sin\theta_p \cos\theta_y + \sin\theta_r \sin\theta_y$$
$$a_{32} = \cos\theta_r \sin\theta_p \sin\theta_y + \sin\theta_r \cos\theta_y$$
$$a_{33} = \cos\theta_r \cos\theta_p$$

Due to the composite rotation, a scatterer at $\vec{S}_1 = [X_1, Y_1, Z_1]$ observed in the (X, Y, Z) coordinate system will rotate to $\vec{S}_2 = [X_2, Y_2, Z_2]$ in the same coordinates. $\vec{S}_1$ and $\vec{S}_2$ is related by the rotation matrix:

$$\vec{S}_2 = \{Rot(\theta_r, \theta_p, \theta_y)\vec{S}_1^T\}^T \tag{5.21}$$

where T denotes the transpose of the vector. In general, the composite translation and rotation matrix is [21]

$$Trans(T_X, T_Y, T_Z)Rot(\theta_r, \theta_p, \theta_y) = \begin{bmatrix} a_{11} & a_{12} & a_{13} & T_X \\ a_{21} & a_{22} & a_{23} & T_Y \\ a_{31} & a_{32} & a_{33} & T_Z \\ 0 & 0 & 0 & 1 \end{bmatrix}$$

$$\tag{5.22}$$

With the composite translation and rotation matrix, the coordinates of any maneuvering target can be calculated by given rotation angles $(\theta_r, \theta_p, \theta_y)$ and translations (T_X, T_Y, T_Z).

5.4.2 Radar Imaging of Maneuvering Target Using Time-Frequency-Based Image Formation

To demonstrate radar imaging of maneuvering targets, we first use simulated radar data, and then apply the time-frequency-based image formation to measured radar data.

In the simulation, the radar is assumed to be operating in X-band at a center frequency of f_0 = 9,000 MHz and transmits a stepped-frequency waveform. Any other waveform, such as linear frequency-modulated and chirp-pulse waveforms, can also be used for the time-frequency-based image formation. A total of M = 64 stepped frequencies are used with a frequency step of 8 MHz to cover a 500 MHz bandwidth or achieve 0.29m range resolution. Each pulse only transmits one carrier frequency wave. After transmitting a group of 64 pulses at 64 stepped frequencies called a burst, the radar transmits another burst. In our simulation, the PRF is 20,000 pulses/sec, which is at least 64 times higher than the burst repetition frequency to generate an image covering the entire target. The image observation time should be long enough to achieve the desired cross-range resolution. In the simulation, coherent image processing time $T = MN/PRF$ = 1.64 sec with N = 512 samples of the time history series is used. Thus, the radar image consists of 64 range-cells and 512 Doppler frequencies or cross-range cells.

An aircraft (MIG-25) is simulated in terms of 2D reflectivity density function $\rho(x, y)$ characterized by 120 point-scatterers having equal reflectivity. These 120 point-scatterers are distributed along the edge of the 2D shape of the aircraft. The simplified point-scatterer model is very simple compared to the electromagnetic prediction code simulation such as the X-patch. Although the point-scatterers do not represent the actual distribution of the reflectivity, it is convenient for displaying the shape of the formatted image of the target. It is good enough for testing and comparing different motion compensation and image formation algorithms.

The aircraft is initially located at a range of 3,500m and has a fast rotation rate of 10 degrees/sec, which is much higher than the normal rotation rate of producing a clear image of a target. We assume that target's translation motion can be perfectly compensated. However, due to the fast rotation and relatively longer image observation time, even after standard motion compensation, the uncompensated phase error is still large. Thus, the formed image by using the Fourier transform is still blurred as shown in Figure 5.6(a).

When a target has a fast rotational motion, polar reformatting is usually desired [1, 2]. This can eliminate individual scatterers from drifting through their range cells, and allow the Fourier transform to be used properly. However, to perform polar reformatting, the knowledge of initial kinematic parameters of the target is required. In addition, resampling and polar-to-rectangular reformation increases the computational complexity of the image formation process.

With the time-frequency-based image formation, at each time the range and the Doppler frequency shift of each scatterer can be determined. Thus,

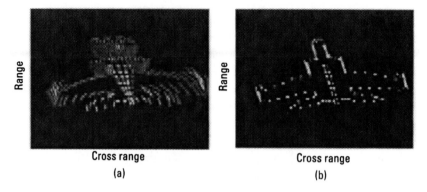

Figure 5.6 (a) Image of a simulated MIG-25 reconstructed with the Fourier-based image formation, and (b) image frame 7 reconstructed with the time-frequency-based image formation.

without knowing the initial kinematic parameters and resampling the data, a blurred Fourier image due to smeared Fourier spectrum will become a sequence of clear range and instantaneous Doppler images.

In principle, any time-frequency transform can be used to replace the Fourier transform for radar image formation. However, a desired time-frequency transform should satisfy the following requirements: (1) it should have high resolution in both the time and frequency domains, and (2) it should accurately reflect the instantaneous frequencies of the analyzed signal.

As discussed in Chapter 2, time-frequency transforms include linear transforms such as the STFT, and bilinear transforms such as the WVD. The joint time-frequency resolution of the STFT is limited by the uncertainty principle. With a time-limited window function, the resolution of the STFT is determined by the window size. There is a trade-off between the time-resolution and the frequency resolution. A larger window has higher frequency resolution but lower time resolution; a smaller window has lower frequency resolution but higher time resolution.

Unlike the STFT, in which the time and frequency resolution is determined by the selection of the short-time window function, there is no short-time window involved in the WVD. The WVD not only has a higher frequency resolution close to that of the full-size windowed Fourier transform, but also provides a higher time resolution. Because of the high resolution and the accuracy of the time-frequency representation, the WVD can be a candidate for time-frequency-based image formation. However, there is cross-term interference associated with the WVD. When the signal contains more than one component, the WVD will generate cross-term interference between components that occurs at spurious locations of the time-frequency plane. The cross-term possesses a limited energy that reflects the correlation between the two related terms and is highly oscillatory. Although the cross-term has a limited contribution to signal energy, it often obscures the useful time-varying spectrum. To reduce the cross-term interference, the filtered WVD can be used to preserve the useful properties of the time-frequency transform with slightly reduced time-frequency resolution and largely reduced cross-term interference. The WVD with linear low-pass filter is characterized as a Cohen's class described in Chapter 2, such as Choi-Williams distribution [5]; and the distribution with a nonlinear low-pass filter is characterized as the TFDS [6] as described in Chapter 2.

The TFDS can have higher resolution and lower cross-term interference depending on its order. When the zero-order is selected, the TFDS is equivalent to the spectrogram of the STFT, and as the order goes to infinity, the TFDS converges to the WVD. In most applications, the order may be selected to be 3 or 4.

As a comparison of the time-frequency energy concentration, the instantaneous frequency and the instantaneous bandwidth for the STFT, the TFDS and the WVD, the WVD has highest time-frequency energy concentration or lowest instantaneous bandwidth, and the instantaneous frequency accurately reflects the true instantaneous frequency of the signal. Depending on the order of the distribution, the TFDS has slightly lower time-frequency energy concentration than the WVD, and can also accurately reflect the true instantaneous frequencies of the signal. But the STFT has lower time-frequency energy concentration and a deviation from the true instantaneous frequencies. In the example described in [17], the instantaneous bandwidth in normalized frequency is 0.007 for the WVD, 0.012 for the fourth-order TFDS, and 0.03 for the STFT. Thus, the time-frequency energy concentration of the STFT is about 4.3 times lower than that of the WVD and about 2.4 times lower than that of the fourth-order TFDS.

Since high time-frequency energy concentration and low cross-term interference are desired for the time-frequency-based image formation, in our simulation we choose the TFDS for its higher time-frequency energy concentration, lower cross-term interferences, and easier implementation.

Figure 5.6(b) shows the image frame 7 from the sequence of 16 frames using the time-frequency-based image formation. The blurred image caused by the target's fast rotation is now refocused without applying the polar reformatting. Because the time-varying spectrum is well represented, the smeared Fourier image is resolved into a sequence of time-varying range and instantaneous Doppler images. These images not only have superior resolution, but also show the Doppler change from one image frame to another and range walk from time to time [17, 18].

To demonstrate the effectiveness of the time-frequency-based image formation for measured radar data, we apply both the Fourier-based and the time-frequency-based image formation to a set of measured radar data returned from a commercial airplane. The radar parameters are about the same as described in the simulated example. The data is also formed as a 64×512 complex I and Q array matrix. After applying the standard motion compensation algorithm, the image formed by the Fourier-based image formation is still blurred as shown in Figure 5.7(a). This is because of the target's fast maneuvering during the entire coherent processing interval. Without knowing its initial kinematic parameters and without resampling the data, a simple way to form an image of the maneuvering target is to use a short-windowed or subaperture data. Because of the short time window, Fourier transform can be adequately applied. However, the Doppler or the cross-range resolution of the reconstructed image is lower because of the short-window as we discuss earlier in Chapter 2. Here, we apply the TFDS described in Chapter 2 to the measured data. The image formed by the time-frequency-based image formation shows higher Doppler or cross-range resolution as seen in Figure 5.7(b). The nose, the wings, the wingtips, the fuselage, the engines, and the tail of the aircraft can be identified clearly. In the upper center of Figure 5.7 the ground truth of the airplane is given for comparison.

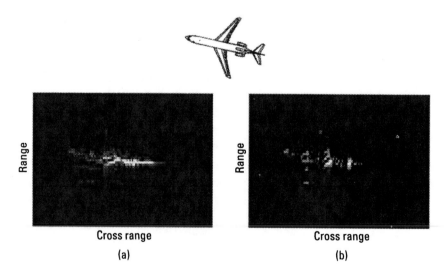

Figure 5.7 Image of an aircraft: (a) reconstructed with the Fourier-based image formation, and (b) reconstructed with the time-frequency-based image formation.

The same time-frequency-based image formation can also be applied to SAR data. In [22], SAR image generated from the subaperture processing and the WVD-based processing is discussed, and the benefit of the time-frequency analysis to SAR data is discussed.

5.5 Radar Imaging of Multiple Targets

Radar imaging of multiple moving targets is an important issue, especially when targets are rotating or maneuvering, such as multiple aircraft in formation flying within the same antenna beam, close to each other, and moving with different velocities or in different directions. The conventional radar imaging algorithms, which work well for single moving targets, cannot be directly applied to the multiple-target environment. Returned signals from these targets are overlapped in time. By simply applying the standard motion compensation algorithm, the phase correction function calculated from the Doppler history derived from a group target cannot compensate the phase error for each individual target. It usually compensates the phase error for one target but induces phase errors for others. Thus, multiple moving targets cannot be resolved, and each individual target cannot be clearly imaged.

To address the issue of multiple targets, we will discuss multitarget resolution in Section 5.5.1. When each individual target has its own radial velocity or Doppler history that is different from others, the time-frequency transform can be used to image each individual target [23–25]. In Section 5.5.2 we discuss how time-frequency-based phase compensation can be applied to imaging of multiple targets. In Section 5.5.3 we extend the time-frequency-based image formation algorithm to radar imaging of multiple targets. Targets can be either point-targets or extended targets, and can have either translational motion or rotational motion and maneuvering [17, 23].

5.5.1 Multiple-Target Resolution Analysis

When a number of targets are within the same radar antenna beam, the returned signal from L targets can be expressed as the summation of the returned signals from individual targets described in (5.12):

$$s_\Sigma(t) = \sum_{k=1}^{L} \int\int\int \rho_k(x, y, z)$$

$$\exp\left\{-j2\pi\left[\frac{2f_0}{c}R_k(t) + xf_{kx}(t) - yf_{ky}(t)\right]\right\}dxdydz \qquad (5.23)$$

where $\rho_k(x, y, z)$ is the reflectivity function at (x, y, z) in the kth target, $f_{kx} = \dfrac{2f_0}{c} \cos \theta_k(t)$ and $f_{ky} = \dfrac{2f_0}{c} \sin \theta_k(t)$ are the associated components of spatial frequencies, and $R_k(t)$ and $\theta_k(t)$ are the range and the rotation angle of the kth target, respectively.

The phase of the returned signal from the kth target is

$$\Phi_{rk}(r_t) = \frac{4\pi f_0}{c} [R_k(t) + x \cos \theta_k(t) - y \sin \theta_k(t)] \tag{5.24}$$

Thus, the Doppler frequency shift due to translational motion is

$$f_{D_{Trans},k} = \frac{2f_0}{c} V_{R,k} = \frac{2}{\lambda} V_{R,k} \tag{5.25}$$

and the Doppler frequency shift due to rotational motion is

$$f_{D_{Rot},k} = \frac{2f_0}{c} [-(x \sin \theta_{0k} + y \cos \theta_{0k})\Omega_k - (x \cos \theta_{0k} - y \sin \theta_{0k})\Omega_k^2 t] \tag{5.26}$$

where $V_{R,k}$ is the initial radial velocity, θ_{0k} is the initial rotation angle, and Ω_k is the angular velocity of the kth target.

When multiple targets are close to each other and, thus, cannot be separated in range, the only approach to separating these targets is to utilize their Doppler frequency differences.

Figure 5.8(a) illustrates two targets flying along a straight-line flight path with the same velocity V. Assume that the distance between the two aircraft is d, and the range from the radar to the midpoint between the two aircraft is R. The angle between the direction of the flight path and the LOS from the radar to the targets 1 and 2 are ϕ_1 and ϕ_2, respectively. The point, whose corresponding angle is $\phi = (\phi_1 + \phi_2)/2$, should be located near the midpoint between the two targets. Therefore, the Doppler difference between the two targets becomes

$$\Delta f_{D_{12}} = f_{D_{Trans},1} - f_{D_{Trans},2} = \frac{2V}{\lambda} (\cos \phi_1 - \cos \phi_2) \tag{5.27}$$

$$= -\frac{2V}{\lambda} \left(2 \sin \phi \sin \frac{\phi_1 - \phi_2}{2} \right) = -\frac{2V}{\lambda} \frac{d}{R} \sin \phi \sin \phi_2$$

where we use the equation

$$\sin\frac{\phi_1 - \phi_2}{2} = \frac{d}{R}\frac{\sin\phi_2}{2}$$

If $\phi \approx \phi_2$, we have

$$\Delta f_{D_{12}} \approx -\frac{2V}{\lambda}\frac{d}{R}\sin^2\phi \qquad (5.28)$$

Let us examine some instances in which multiple targets can be separated. Assuming the radar is operating at X-band ($\lambda = 0.033$m), and $V = 60$ m/sec, $d = 20$m, and $R = 20,000$m, the Doppler difference, as a function of the angle ϕ, at $\phi = 90$ and 270 degrees, reaches a maximum value of 3.5 Hz as shown in Figure 5.8(a). To distinguish two targets, a coherent processing time of 0.28 sec is needed. If the angle ϕ is away from 90 or 270 degrees, the Doppler difference becomes smaller. Thus, a longer coherent processing time, which may cause image blurring, is required to resolve two targets. When the angle ϕ approaches 0 or 180 degrees, it is impossible to distinguish these two targets.

Figure 5.8(b) illustrates another example of two targets lined up and flying in the same direction. In this case, the Doppler difference between the two targets becomes

$$\Delta f_{D_{12}} = f_{D_{Trans},1} - f_{D_{Trans},2} = \frac{2V}{\lambda}(\cos\phi_1 - \cos\phi_2)$$

$$= -\frac{2V}{\lambda}\left(2\sin\frac{\phi_1 + \phi_2}{2}\sin\frac{\phi_1 - \phi_2}{2}\right) \qquad (5.29)$$

$$= -\frac{2V}{\lambda}\left[2\sin\left(\frac{d}{2R} + \phi_2\right)\sin\left(\frac{d}{2R}\right)\right]$$

If d/R is very small, then

$$\Delta f_{D_{12}} \approx -\frac{2V}{\lambda}\frac{d}{R}\sin\left(\frac{d}{2R} + \phi_2\right) \qquad (5.30)$$

For the same parameters V, d, and R given above, a function of the angle ϕ_2 the Doppler difference at $\phi_2 = 90$ and 270 degrees reaches the same

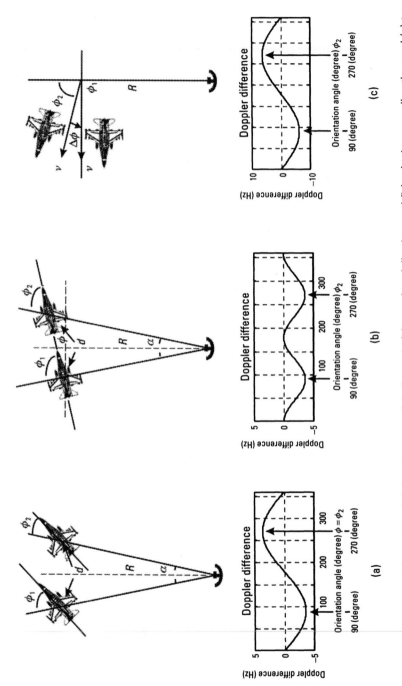

Figure 5.8 Geometry of (a) two aircraft flying along a straight-line path; (b) two aircraft lined-up and flying in the same direction; and (c) two aircraft at the same range but moving in different directions, and their corresponding Doppler difference as a function of the angle.

maximum value of 3.5 Hz as shown in Figure 5.8(b). When the angle ϕ_2 approaches 0 or 180 degrees, the Doppler difference approaches zero; thus, it is impossible to distinguish these two targets.

In the case of multiple targets at the same range but moving in different directions as shown in Figure 5.8(c), the Doppler difference between two targets with angle ϕ_1 and ϕ_2 becomes

$$\Delta f_{D_{12}} = f_{D_1} - f_{D_2} = \frac{2V}{\lambda}(\cos\phi_1 - \cos\phi_2) \tag{5.31}$$

$$= -\frac{2V}{\lambda}\left(2\sin\left(\frac{\Delta\phi}{2} + \phi_2\right)\sin\frac{\Delta\phi}{2}\right)$$

If $\phi_2 = 180$ degrees, then

$$\Delta f_{D_{12}} = \frac{2V}{\lambda}\left(2\sin^2\frac{\Delta\phi}{2}\right) \tag{5.32}$$

where $\Delta\phi$ is the difference between the direction angles of the two targets.

From (5.31), if $V = 60$ m/sec, $R = 20,000$m, and $\Delta\phi = 0.1$ degree, as a function of the angle ϕ_2, at $\phi_2 = 90$ and 270 degrees, the Doppler difference reaches a maximum value of 6.35 Hz as shown in Figure 5.8(c). To distinguish two targets, a coherent processing time of 0.16 sec is enough. When the angle ϕ_2 approaches 0 or 180 degrees, the Doppler difference approaches zero, and it is also impossible to distinguish these two targets.

From the first two examples, we can ascertain that when targets are flying in the same direction, the Doppler difference between targets is relatively small. With a longer coherent processing time, multiple targets may be resolved. When, however, targets are flying in different directions as shown in the third example, the Doppler difference is relatively large depending on the angle difference of their flying directions.

If targets have rotational motion or different velocities, they will have different Doppler histories. By using conventional motion compensation algorithms, an image of multiple targets becomes smeared. In these cases, time-frequency algorithms may help for imaging of multiple targets.

5.5.2 Time-Frequency-Based Phase Compensation for Multiple Targets

When a number of targets are within the same antenna beam, close to each other in range, and moving with different velocities or in different directions, each individual target has its own Doppler history. Although these targets are difficult to separate in range, the difference in Doppler histories can be used to resolve multiple targets. An approach to separate different Doppler histories is to apply the time-frequency transform to the time history series data at range cells where targets are located [23–25]. From each individual time-frequency distribution that associates with each specific target, Doppler history of that target is obtained. By taking the time-integral of the Doppler history, the phase history function associated with that target can be found. Then, multiplying the conjugate of the phase history function to the raw radar data, the phase function associated with the specific target will be compensated and its Doppler frequency shift becomes time-invariant. Thus, the Fourier transform can be adequately applied to reconstruct the image of that specific target. To image multiple targets, the above procedure of phase compensation must be applied to each of the targets separately as illustrated in Figure 5.9 where three targets are imaged using three different phase functions.

If, however, a target has rotational motions, the phase compensation procedure described above may not work well and, thus, images of multiple moving targets may not be well focused.

Here is an example of two targets, each of which has a circular motion around its own center point as shown in Figure 5.10(a). Radar is located at $(X = 0, Y = 0)$, and the two targets start their circular motion from a same

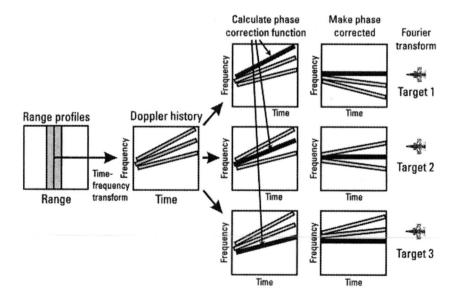

Figure 5.9 Conventional time-frequency approach to multiple targets.

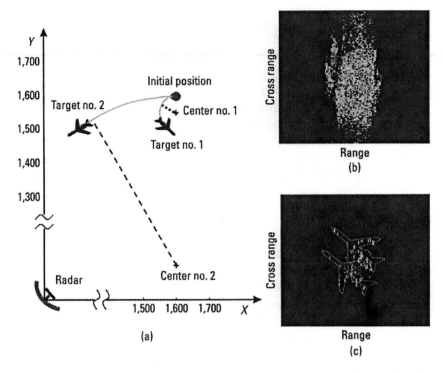

Figure 5.10 (a) Trajectories of two simulated targets with circular motion; (b) radar imaging of multiple targets using time-frequency-based phase compensation and Fourier transform; and (c) radar imaging of multiple targets using time-frequency transform.

initial point at $(X = 1,600m, Y = 1,600m)$. Target 1 has a circular motion around a center at $(X = 1,600m, Y = 1,550m)$ and with a rotation rate 0.04 rad/sec or 2.29 degrees/sec. Target 2 has a rotation rate of 0.02 rad/sec or 1.15 degrees/sec around a center at $(X = 1,600m, Y = 1,100m)$. Because of the rotation, each scatterer on the target has its own unique combined translational and rotational motions, and therefore, has its own Doppler history. In this case, the simple phase compensation algorithm cannot remove Doppler drifts of all the individual scatterers. Thus, the image of each individual target still can be smeared, and multiple targets cannot be seen as shown in Figure 5.10(b).

5.5.3 Time-Frequency-Based Image Formation for Radar Imaging of Multiple Targets

In cases where the phase compensation algorithm does not work well for multiple rotating or maneuvering targets, we can apply the time-frequency-based image formation. As discussed in Section 5.4.2, the time-frequency-based image formation works well in cases where each individual scatterer of the target has its own range and Doppler history, such as scatterers in multiple targets with rotations. When we apply the time-frequency-based image formation, multiple targets can be identified from a sequence of range and instantaneous Doppler image frames. The target can be either a point target or an extended target; and the target's motion can be either translational or maneuvering. Figure 5.10(c) shows the two separated aircraft reconstructed by the time-frequency-based image formation compared with the smeared one in Figure 5.10(b) where the phase compensation algorithm is applied.

5.6 Summary

In this chapter, we introduced a new radar image formation called the time-frequency-based image formation. Compared to the Fourier-based image formation, the time-frequency approach works well for maneuvering targets. In principle, any time-frequency transform can be used for radar image formation. We use the TFDS because of its higher resolution and lower cross-term interference. However, any other time-frequency transform with higher resolution and lower cross-term interference can also be used. The time-frequency-based image formation can be used for either ISAR or SAR image formation and works well in either single-target or multiple-target environments.

References

[1] Wehner, D. R., *High-Resolution Radar, Second Edition*, Norwood, MA: Artech House, 1994.

[2] Carrara, W. G., R. S. Goodman, and R. M. Majewski, *Spotlight Synthetic Aperture Radar—Signal Processing Algorithms*, Norwood, MA: Artech House, 1995.

[3] Harger, R. O., *Synthetic Aperture Radar System*, New York: Academic Press, 1970.

[4] Son, J. S., G. Thomas, and B. Flores, *Range-Doppler Radar Imaging and Motion Compensation*, Norwood, MA: Artech House, 2000.

[5] Cohen, L., *Time-Frequency Analysis*, Englewood Cliffs, NJ: Prentice Hall, 1995.

[6] Qian, S., and D. Chen, *Joint Time-Frequency Analysis—Methods and Applications*, Englewood Cliffs, NJ: Prentice Hall, 1996.

[7] Gabor, D., "Theory of Communication," *J. IEE*, Vol. 93, No. III, 1946, pp. 429–457.

[8] Bastiaans, M. J., "On the Sliding-Window Representation in Digital Signal Processing," *IEEE Trans. Acoust., Speech, Signal Processing*, Vol. ASSP-33, No. 4, 1985, pp. 868–873.

[9] Ausherman, D. A., et al., "Developments in Radar Imaging," *IEEE Trans. Aerospace and Electronic Systems*, Vol. 20, No. 4, 1984, pp. 363–400.

[10] Prickett, M. J., and C. C. Chen, "Principles of Inverse Synthetic Aperture Radar (ISAR) Imaging," *IEEE 1980 EASCON*, 1980, pp. 340–345.

[11] Chen, C. C., and H. C. Andrews, "Target Motion Induced Radar Imaging," *IEEE Trans. Aerospace and Electronic Systems*, Vol. 16, No. 1, 1980, pp. 2–14.

[12] Walker, J., "Range-Doppler Imaging of Rotating Objects," *IEEE Trans. Aerospace and Electronic Systems*, Vol. 16, No. 1, 1980, pp. 23–52.

[13] Wahl, D. E., et al., "Phase Gradient Autofocus—A Robust Tool for High Resolution SAR Phase Correction," *IEEE Trans. Aerospace and Electronic Systems*, Vol. 30, No. 3, 1994, pp. 827–834.

[14] Kirk, J. C., "Motion Compensation for Synthetic Aperture Radar," *IEEE Trans. Aerospace and Electronic Systems*, Vol. 11, 1975, pp. 338–348.

[15] Wu, H., et al., "Translational Motion Compensation in ISAR Image Processing," *IEEE Trans. Image Processing*, Vol. 14, No. 11, 1995, pp. 1561–1571.

[16] Rihaczek, A. W., and S. J. Hershkowitz, *Radar Resolution and Complex-Image Analysis*, Norwood, MA: Artech House, 1996.

[17] Chen, V. C., and S. Qian, "Joint Time-Frequency Transform for Radar Range-Doppler Imaging," *IEEE Trans. Aerospace and Electronic Systems*, Vol. 34, No. 2, 1998, pp. 486–499.

[18] Chen, V. C., "Reconstruction of Inverse Synthetic Aperture Radar Image Using Adaptive Time-Frequency Wavelet Transform," *SPIE Proc. on Wavelet Applications*, Vol. 2491, 1995, pp. 373–386.

[19] Chen, V. C., and H. Ling, "Joint Time-Frequency Analysis for Radar Signal and Image Processing," *IEEE Signal Processing Magazine*, Vol. 16, No. 2, 1999, pp. 81–93.

[20] Sparr, T., S. Hamran, and E. Korsbakken, "Estimation and Correction of Complex Target Motion Effects in Inverse Synthetic Aperture Imaging of Aircraft," *Proc. Of IEEE Intl. Radar Conference*, 2000, pp. 457–461.

[21] Chen, V. C., and W. J. Miceli, "Time-Varying Spectral Analysis for Radar Imaging of Maneuvering Targets," *IEE Proceedings—Radar, Sonar and Navig.*, Vol. 145, No. 5, 1998, pp. 262–268.

[22] Fiedler, R., and R. Jansen, "Joint Time-Frequency Analysis of SAR Data," *Proceedings of the Tenth IEEE Workshop on Statistical Signal and Array Processing*, 2000, pp. 480–484.

[23] Chen, V. C., and Z. Z. Lu, "Radar Imaging of Multiple Moving Targets," *SPIE Proc. On Radar Processing, Technology, and Applications II*, Vol. 3161, 1997, pp. 102–112.

[24] Wu, X., and Z. Zhu, "Multiple Moving Target Resolution and Imaging Based on ISAR Principle," *Proc. of IEEE 1995 National Aerospace and Electronic Conference*, 1995, pp. 982–987.

[25] Wang, A., Y. Mao, and Z. Chen, "Imaging of Multi-Targets with ISAR Based on the Time-Frequency Distribution," *ICASSP*, 1994, pp. V-173–176.

6

Motion Compensation in ISAR Imaging Using Time-Frequency Techniques

As discussed in the last chapter, radar imaging is a process of mapping the electromagnetic reflectivity of a target from multiple-frequency, multiple-aspect data. Frequency diversity can be fairly easily achieved by the built-in bandwidth of the radar sensor. Angular diversity, on the other hand, must be acquired through the relative movement between the target and the sensor. In SAR, the sensor is moved around the target to acquire the necessary angular data, while in ISAR imaging, the stationary sensor collects multiple aspect data through target movement. For instance, a ground-based radar observing an in-flight aircraft over a sufficient time interval can collect data needed to form an ISAR image.

One of the main challenges in ISAR image formation is the unknown nature of the target motion. Ideally, if the target has no translation motion and only uniform rotational motion, then a simple Fourier transform process would bring a set of range profiles collected over a given dwell time (i.e., the coherent processing interval) into a focused 2D image. However, this is never true in real-world ISAR imaging scenarios, as the target being imaged is often engaged in complicated maneuvers that combine translation and rotational motions. Therefore a process called motion compensation must be counted on to form a focused image of the target. This is a blind process since the radar data is the only available information. Two basic assumptions are usually used to aid us in this difficult task, namely, the point-scatterer model and the rigid body assumption.

123

In this chapter, we examine the use of time-frequency analysis for achieving ISAR motion compensation. In Section 6.1, we review some of the existing motion compensation algorithms for translation motion and rotational motion removal. In Section 6.2, we introduce an adaptive joint time-frequency (AJTF) procedure for extracting the phase of a prominent point-scatterer on the target from the radar data. We show how the extracted phase information can be used in conjunction with the prominent point processing (PPP) model to achieve motion compensation. In Section 6.3, we illustrate this algorithm using both simulation and measurement data. In Section 6.4, we discuss the case when the rotational motion of the target is not confined to a 2D plane.

6.1 Motion Compensation Algorithms

In real-world ISAR data collection, the target being imaged is usually engaged in complex maneuvers that combine translation and rotational motions (see Figure 6.1). Unless a good motion compensation algorithm is implemented, serious blurring can result in the ISAR image formed by the Fourier transform, which assumes that all the point-scatterers in a range cell behave with linear phase across different pulses. A motion compensation algorithm typically consists of two parts, range alignment and Doppler tracking. Among all

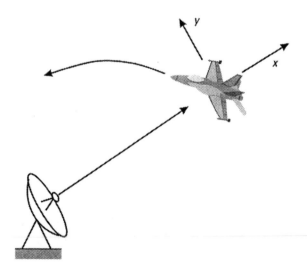

Figure 6.1 Complex target motion during ISAR image collection can be considered a combination of translation and rotational motions with respect to the radar.

the existing motion compensation algorithms, the manner in which range alignment is performed is fairly standard. It is accomplished by tracking the movement of a reference point (such as a prominent peak or the centroid of the range profile) across multiple pulses and fitting it to a low-order polynomial [1]. The result of such coarse range alignment is illustrated in Figure 6.2(a). The coarse range alignment allows a point-scatterer to be sorted into the same range bin across all the pulses. However, the accuracy of the alignment is limited by the range resolution, which is typically tens of centimeters. This is not sufficient to overcome the phase errors measured in terms of the radar operating wavelength, which is typically a few centimeters. Consequently, Doppler tracking must be carried out to align the phase. There are many different schemes to perform Doppler tracking, including the sub-aperture approach [2–4], the cross-range centroid tracking approach

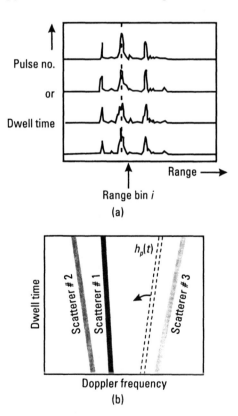

Figure 6.2 (a) Range profile versus pulse number after coarse range alignment; and (b) time-frequency representation of the signal in range bin *i* and the search procedure for the best basis *hp(t)*.

[5], and the phase gradient autofocus (PGA) technique [6]. These methods consider the Doppler frequency shifts of the target as a whole, and apply the same correction vector to all of the scatterers in the image. This is effective when translation motion is dominant.

When the coherent processing interval is long or when the target exhibits fast maneuvers, the phase error due to the nonuniform rotational motion is often not negligible and must also be properly compensated. One useful scheme to carry out this operation is the multiple PPP algorithm [7, 8]. The basic idea of the multiple PPP algorithm is to track one or more point-scatterers in the image in order to extract the motion parameters. Once the motion parameters are known, compensation of both translation and rotational motions can be achieved. The main challenge in applying the multiple PPP algorithm is the selection of the prominent point-scatterers, which should ideally be well isolated in their respective range cells. In practice, such kinds of point-scatterers may be difficult to pinpoint or not available at all.

As was described in Chapter 5, time-frequency analysis is an attractive way to address the Doppler tracking issue in motion compensation [9, 10]. Specifically, in [10], it was shown that by applying the TFDS [11] described in Section 2.2.3 in place of the Fourier transform engine, the ISAR image can be effectively examined at each dwell time instance, thus eliminating range drift and Doppler smearing. Unfortunately, the TFDS is based on the WVD and does not preserve the phase information of the original image. Such information may be important for subsequent feature extraction or feature matching operations in target identification. Furthermore, the Doppler resolution achievable in this manner is still less than that offered by the total dwell time interval of the original data. In the next section, we describe a time-frequency-based procedure for achieving both translation and rotational motion compensation [12, 13]. An adaptive procedure extended from the adaptive spectrogram [14] discussed in Section 2.1.3 is used to select and extract the phase of multiple prominent point-scatterers on the target. The extracted phase is then coupled with the multiple PPP model to eliminate the undesirable motion errors in the original radar data. In this manner, the phase of the focused image is preserved and the Doppler resolution offered by the full coherent processing interval can be achieved.

6.2 Time-Frequency-Based Motion Compensation

We first introduce the standard motion model, which has been alluded to in Chapter 5, for describing the radar signal. The model is based on the rigid-body, point-scatterer assumption. The target is assumed to contain both translation and rotational motions. However, coarse range alignment is assumed to have been applied to the data so that the point-scatterers have been sorted into their respective range cells [Figure 2(a)]. The target is assumed to have a dominant rotational axis during the imaging interval (see Section 6.4 for discussions on the more general case when this assumption does not hold). We can then express the radar received signal within a particular range cell at x as

$$s_R(t)\big|_x = \sum_{k=1}^{N_k} A_k \exp\left\{-j\frac{4\pi f_0}{c}[R(t) + x\cos\theta(t) - y_k\sin\theta(t)]\right\} \quad (6.1)$$

where the x-axis is the down-range direction. In (6.1), f_0 is the center frequency of the radar and t denotes the dwell time (which is proportional to the pulse number). N_k is the number of point-scatterers within the range cell and all of them share the same down range position x. A_k and y_k are respectively the magnitude and cross range position of the kth point-scatterer. $R(t)$ describes the residual uncompensated translation displacement and $\theta(t)$ is the rotation angle as a function of dwell time (measured from the center of the imaging interval). As shown in (5.2) and (5.3), we expand $R(t)$ and $\theta(t)$ into their respective Taylor series as follows:

$$R(t) = R_0 + V_R t + \frac{1}{2}a_R t^2 + \ldots \quad (6.2)$$

$$\theta(t) = \Omega t + \frac{1}{2}\gamma t^2 + \ldots \quad (6.3)$$

By substituting these expressions into (6.1) and taking the leading terms, we obtain

$$s_R(t)\big|_x = \sum_{k=1}^{N_k} A_k \exp\left\{-j\frac{4\pi f_0}{c}[(R_0 + x) + (V_R + \Omega y_k)t \quad (6.4)\right.$$
$$\left. + \frac{1}{2}(a_R - \Omega^2 x + \gamma y_k)t^2 + \ldots]\right\}$$

The constant phase term is not relevant to the imaging process and can be ignored. Note that if $R(t) = R_0$, $\theta(t) = \Omega t$ and $(\Omega t)^2$ is negligible, phase function is a pure linear function of time. This is the ideal case when a simple Fourier transform will focus the point-scatterers to their respective y_k positions in the cross-range dimension. The resulting cross-range resolution is inversely proportional to the total imaging time interval. However, when the quadratic (and higher-order) term in the phase function is significant, simple Fourier processing will lead to image blurring. The goal of the motion compensation algorithm is to estimate and eliminate the quadratic (and higher-order) phase terms.

6.2.1 Estimating Phase Using Adaptive Time-Frequency Projection

We observe that the quadratic phase term in the uncompensated radar signal under consideration behaves like a linear chirp in the time-frequency plane. As shown in Figure 6.2(b), the dwell-time and Doppler-frequency trajectories of the point-scatterers within a given range cell are straight lines. The displacement and slope of each line are related respectively to the linear and quadratic coefficients of their phase function. The task at hand is to determine these coefficients for the dominant point-scatterer within the range cell. We consider here a time-frequency procedure extended from the adaptive Gaussian representation discussed in Section 2.1.3. More general algorithms for the decomposition of signals using chirp basis can be found in [15, 16]. In our present problem, the point-scatterers have already been aligned in range and it is not necessary to take into account the amplitude variation in the basis function. We do, however, add an additional quadratic (and possibly higher-order) phase term in the basis function as follows:

$$h_p(t) = \exp\left[-j2\pi\left(f_{D_0}t + \frac{1}{2}f_{D_1}t^2 + \ldots\right)\right] \tag{6.5}$$

The above set of bases can be thought of as a collection of unit chirps, each with a different displacement and chirp slope [shown as a dashed line in Figure 6.2(b)]. Next, we carry out a search procedure (as in the adaptive spectrogram) to determine the best basis by projecting the radar signal onto all possible bases of the form (6.5). We search the parameters $(f_{D_0}, f_{D_1}, \ldots)$ that satisfy

$$<f_{D_0}, f_{D_1}, \ldots> = \arg\max\left|\int s_R(t)\big|_x h_p^*(t)dt\right| \tag{6.6}$$

Equation (6.6) implies that the parameters in the phase function are estimated to give the maximum projection from the radar data onto the basis function. In the adaptive spectrogram, this procedure is iterated to try to parameterize the total signal. However, in the present application, only the strongest point-scatterer within a range cell is searched. Because of the rigid body assumption, the motion parameters in (6.2) are carried by every point-scatterer on the target. We choose to use only the dominant scatterer in a range cell in order to avoid estimation errors for the weaker scatterers. The search for the linear coefficient f_{D_0} can be accomplished by using the fast Fourier transform. Then only a 1D search is required to find f_{D_1}, the quadratic coefficient. This procedure can also be extended for cubic and higher-order coefficients, at the expense of more computation time. In terms of performance, the algorithm is equivalent to picking out the strongest line in the time-frequency plane with the full Doppler resolution offered by the total coherent processing interval. Also, this projection algorithm applies even when there is not an isolated, dominant point-scatterer in the range cell.

6.2.2 Motion Error Elimination

With the AJTF procedure for estimating the phase of individual point-scatterers in hand, the next task is to eliminate those quadratic phase terms for all the scatterers. This can be accomplished by the standard multiple PPP algorithm [7, 8]. We consider here the case when $(\Omega t)^2 \ll 1$ so that this term can be neglected in the phase. This is usually a good approximation for high-frequency radars (X-band and above), as the angular window needed to form an image with sufficient cross-range resolution is quite small in absolute terms (i.e., a few degrees). With the approximation, (6.1) becomes

$$s_R(t)\big|_x = \sum_{k=1}^{N_k} A_k \exp\left\{-j\frac{4\pi f_0}{c}\left[(V_R + \Omega y_k)t + \frac{1}{2}(a_R + \gamma y_k)t^2\right]\right\}$$

(6.7)

We observe that the quadratic phase coefficient consists of two terms. The first term, $\left(\frac{1}{2}a_R t^2\right)$, represents the translation motion error and is independent of the cross-range y. The second cross-range-dependent term, $\left(\frac{1}{2}\gamma y_k t^2\right)$, represents the rotational motion error. We first carry out the translation motion compensation by extracting the phase of a prominent point-scatterer located at (x_1, y_1) using the AJTF search. The estimated phase is denoted as

$$\exp\left[-j2\pi\left(f_{D_{01}}t + \frac{1}{2}f_{D_{11}}t^2\right)\right]$$

(6.8)

We then multiply the radar data by the conjugate of (6.8). Based on the model, the phase of an arbitrary point-scatterer at (x_i, y_i) on the target is reduced to

$$\exp\left\{-j\frac{4\pi f_0}{c}\left[(y_i - y_1)\left(\Omega t + \frac{1}{2}\gamma t^2\right)\right]\right\}$$

(6.9)

As can be seen from (6.9), the translation error has been removed from the data and only rotational motion error remains. The reference point (x_1, y_1) serves as the center of the rotational motion.

Next we carry out the rotational motion compensation by extracting the phase of a second prominent point-scatterer at (x_2, y_2):

$$\exp\left[-j2\pi\left(f_{D_{02}}t + \frac{1}{2}f_{D_{12}}t^2\right)\right]$$

(6.10)

By comparing (6.10) to (6.9), we see that the extracted phase gives the desired relationship between the rotation angle and the dwell time (up to a proportionality constant):

$$\theta(t) = \Omega t + \frac{1}{2}\gamma t^2 \propto f_{D_{02}}t + \frac{1}{2}f_{D_{12}}t^2$$

(6.11)

Once this relationship is found, it is possible to reformat the radar data to eliminate the quadratic phase dependence on dwell time. The original radar data is uniformly sampled in dwell time t. We interpolate the data based on (6.11) such that it becomes uniformly sampled in angle θ. After the reformatting, the phase of each point-scatterer is linearly related to the angle and the residual motion due to nonuniform rotation rate is thus removed from the data.

To summarize, by extracting the phase of two prominent point-scatterers using the AJTF projection technique, both the translation and rotational motions can be removed from the radar data. The resulting image is focused since all the point-scatterers have linear phase behaviors. Although we have limited our discussion to quadratic errors, it is straightforward to extend the algorithm to higher-order motion errors. In addition, we have assumed the absence of any range-dependent phase errors [e.g., $(1/2)(\Omega t)^2 x$ in (6.4)], which may appear in wide-angle imaging scenarios. Based on the PPP model, this type of error can also be removed by using three prominent point-scatterers [13].

6.3 Motion Compensation Examples of Simulated and Measured Data

Simulated radar data from a Boeing 727 airplane is used as an example to demonstrate the AJTF motion compensation procedure. The center frequency of the radar is 9 GHz and the bandwidth is 150 MHz. The total number of pulses used to form the image is 256. Figure 6.3(a) shows the ISAR image after range alignment and Doppler centroid focusing. Since Doppler centroid focusing has only limited accuracy and the target contains a significant amount of nonuniform rotational motion, only a portion of the airplane is well focused and the parts away from the focus are seriously blurred. To describe the underlying mechanisms of autofocusing more explicitly, we plot the variation of Doppler frequency versus dwell time in a particular range cell in the time-frequency plane in Figure 6.3(b). The STFT is used to generate the spectrogram. Each line in the spectrogram represents the time-varying Doppler characteristics of a scattering center in the range cell. The displacement of the line represents the coefficient of the linear phase term and its slope represents the coefficient of the quadratic term. It is difficult to accurately extract the desired parameters from such a fuzzy image due to the low resolution of the STFT. Instead, we use the AJTF algorithm to search for the phase of the strongest point-scatterer in this range cell. After multiplying its phase correction to the range-aligned data, an image with the translation motion error removed is obtained and shown in Figure 6.4(a). As can be seen from the spectrogram of Figure 6.4(b), the strongest line has been straightened and shifted to the center of the Doppler frequency axis. However, rotational motion error remains and the other point-scatterers in the range bin still exhibit time-varying Doppler. Thus, a second reference point must be selected to carry out the rotational motion compensation. After extracting its phase via the AJTF procedure, we carry out a reformatting operation such that the radar data becomes uniformly sampled in angle instead of dwell time. As shown in Figure 6.5(a), the

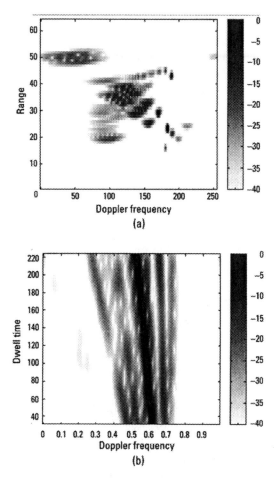

Figure 6.3 (a) ISAR image of simulated Boeing-727 data after range alignment and Doppler centroid focusing; and (b) STFT spectrogram of the signal at a chosen range cell. (*Source:* [13] © 1998 IEEE.)

resulting image is well focused. In the spectrogram of Figure 6.5(b), all the lines are straightened, implying that the phase errors contained in all the point-scatterers have been removed. In Figure 6.6, the image from the simulated data without any added motion errors is shown as a reference of comparison. As we can see, the motion compensated image achieves the same sharpness as the reference image.

The AJTF motion compensation algorithm is next demonstrated using measured ISAR data [17]. The radar data was collected using a ground radar and the target was an aircraft in flight. Figure 6.7(a) shows the coarsely

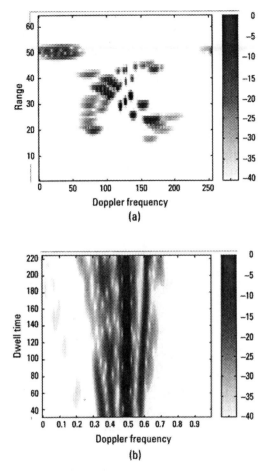

Figure 6.4 (a) ISAR image of simulated Boeing-727 data after translation motion compensation using one prominent point-scatterer; and (b) STFT spectrogram of the signal at the chosen range cell. (*Source:* [13] © 1998 IEEE.)

aligned range profiles over 128 pulses. Figure 6.7(b) shows the resulting ISAR image from taking a series of 1D Fourier transforms across the pulse number (or dwell time). Since significant phase errors still exist in the data, the image is quite blurry in the Doppler dimension. Figure 6.7(c) shows the spectrogram of the strongest range cell. It can be seen that the trajectories of the point-scatterers are tilted. In addition, slight curvature can be noticed in the trajectories. Therefore, we used polynomials of order three (i.e., including linear, quadratic, and cubic terms) in the basis search. We then multiplied the radar data by the conjugate of the best-fit basis to remove

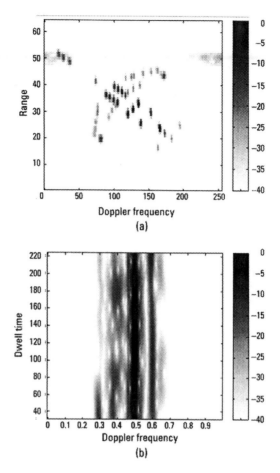

Figure 6.5 (a) ISAR image of simulated Boeing-727 data after rotational motion compensation using a second prominent point-scatterer; and (b) STFT spectrogram of the signal at the chosen range cell. (*Source:* [13] © 1998 IEEE.)

the residual translation motion. Figure 6.8(a) shows the spectrogram of the same range bin as Figure 6.7(c). The motion compensation has successfully straightened the time-frequency trajectories of the point-scatterers. Figure 6.8(b) shows the image formed from the motion-compensated data by using Fourier processing. The image is now well focused and the shape of the aircraft is clearly visible. In this image frame, there is negligible rotational motion error, as additional rotational motion compensation did not further improve the image quality. To more objectively evaluate the performance of the blind motion compensation, a truth image was also generated using

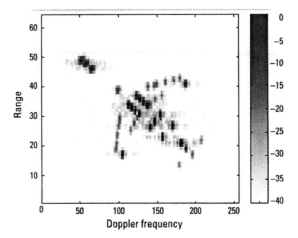

Figure 6.6   Reference ISAR image of simulated Boeing-727 data without any motion error. (*Source:* [13] © 1998 IEEE.)

the known motion of the aircraft. In this case, the aircraft motion was recorded independently using a global positioning system (GPS) and an inertial navigation system (INS) sensor carried on-board the aircraft during the data collection. This data was used to establish the true target motion and compensate the radar data to generate the truth image shown in Figure 6.8(c). By comparing Figure 6.8(b) from the blind motion compensation algorithm to the truth image in Figure 6.8(c), we see that the motion compensation algorithm performed very well in achieving a focused image of the target.

To summarize, the AJTF algorithm provides an automatic means of tracking the phase of one or more point-scatterers in the radar data with high resolution. This algorithm, when combined with the multiple PPP model, is shown to be an effective method of achieving both translation and rotational motion compensation for ISAR imaging. The resulting complex I and Q data is free of phase errors and leads to a well-focused ISAR image. This algorithm has been tested using both simulation and measured data sets.

6.4   Presence of 3D Target Motion

A fundamental assumption in the motion model shown in (6.1) is that the rotational motion of the target is confined to a 2D plane during the coherent

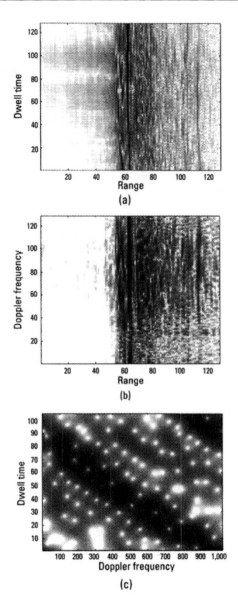

Figure 6.7 ISAR image formation from measured data: (a) range profiles versus dwell time after range alignment; (b) resulting ISAR image; and (c) spectrogram of the strongest range cell.

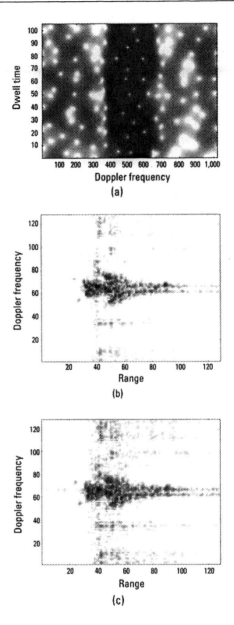

Figure 6.8 ISAR image formation from measured data: (a) spectrogram of the strongest range cell after AJTF motion compensation; (b) resulting motion-compensation image; and (c) truth image formed using actual motion data. (*Source:* [17] © 1999 SPIE.)

processing interval. We shall use the term 2D motion to refer to target rotation of this type. We also saw how under this assumption, the rotational motion can be fully compensated by tracking the phase of a second point-scatterer on the target. However, for aircraft undergoing fast maneuvers or ships on rough seas, the motion of a target may be more chaotic and does not always obey the 2D motion model [17–19]. As a result, the application of a motion compensation algorithm based on the 2D motion model to such intervals can lead to blurry images. In this section, we discuss the case when there exists 3D target motion during the imaging interval (i.e., when the rotational motion is not confined to a 2D plane). Alternatively, we can think of this type of motion as having a time-varying rotational axis during the imaging interval. Figure 6.9(a) shows an image formed using the same data set and the same motion compensation algorithm as those used for Figure 6.8(a), except the data is from a different time interval. The severe image blurring can be attributed to 3D motion. In Figure 6.9(b), we plot the truth motion data generated from on-board motion sensors. It shows the pose angles (both azimuth and elevation) of the target with respect to the radar during the imaging interval. During the first half of the 128 records, the aircraft undergoes rotation primarily along the elevation dimension, while during the second part it undergoes rotation along the azimuth dimension. If this information were available, we could avoid choosing such an imaging interval. However, we do not usually have access to the attitude data on non-cooperative targets. In such cases, the questions of how to detect the presence of 3D rotational motion and how to pinpoint the "good imaging intervals" where the target obeys the 2D motion model must be addressed. We describe here an algorithm based on the same AJTF engine to detect blindly the presence of 3D motion based on the radar data alone [20, 21].

Let us begin our discussion by considering the 2D and 3D rotational motion models. We shall assume the target rotational angle is small and we can apply the small-angle approximation ($\cos\theta \approx 1$, $\sin\theta \approx \theta$) to (6.1). The 2D model is then given by:

$$s_R(t)\big|_x = \sum_{k=1}^{N_k} A_k \exp\left\{-j\frac{4\pi f_0}{c}[R(t) + x + y_k\theta(t)]\right\} \qquad (6.12)$$

where θ is the rotational angle in the 2D plane. When 3D motion exists, this model must be augmented to account for rotational motion in both θ and the orthogonal ϕ directions. The 3D model is given by

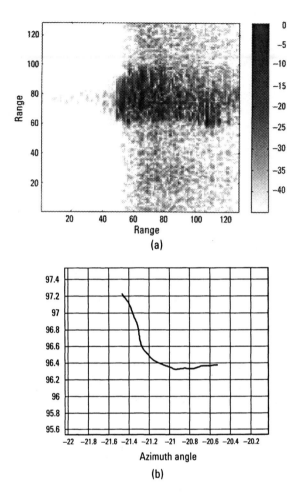

Figure 6.9 (a) ISAR image after AJTF motion compensation for an image frame containing significant 3D rotational motion; and (b) pose angles of the target with respect to the radar during the imaging interval derived from truth motion data. (*Source:* [17] © 1999 IEEE.)

$$s_R(t)\big|_x = \sum_{k=1}^{N_k} A_k \exp\left\{-j\frac{4\pi f_0}{c}[R(t) + x + y_k\theta(t) + z_k\phi(t)]\right\}$$

(6.13)

where another independent angular parameter ϕ is introduced to describe the 3D rotational motion (see Figure 6.10), and (x, y_k, z_k) are the 3D spatial positions of the point-scatterers in the range cell. It can be shown

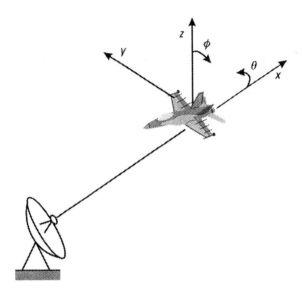

Figure 6.10 Geometry of the ISAR problem involving 3D motion.

that the 3D model reduces to the 2D model under two conditions. The first is when there exists a linear relationship between θ and ϕ [i.e., $\theta(t) = a\phi(t)$]. The second is when the z-dimension of the target is small and the third phase term can be ignored. When these conditions are not met, the full 3D model must be used and any motion compensation algorithm based on the 2D model will not focus the target well. Figure 6.11 shows the result of the point-scatterer simulation. Figure 6.11(a, b) are respectively the assumed rotational motions used in the simulation and the resulting motion-compensated image based on the 2D model. As can be seen from these two figures, the assumed relationship between θ and ϕ is a linear one and the image is easily focused using the 2D motion compensation algorithm described in Section 6.2. Figure 6.11(c, d) show the case when 3D motion is present. After two-point focusing, the two points in the chosen range cells 25 and 57 are indeed focused. Another point-scatterer in range cell 99 is also focused, as it happens to be in the same 2D motion plane as the point-scatterer in range cell 57. However, the rest of the point-scatterers remain unfocused.

Since 2D motion can be represented by a linear relationship between θ and ϕ, we should be able to detect the presence of 3D motion if we can detect the existence of a nonlinear relationship between θ and ϕ. Next, we show how this can be detected from the phase functions. We assume that

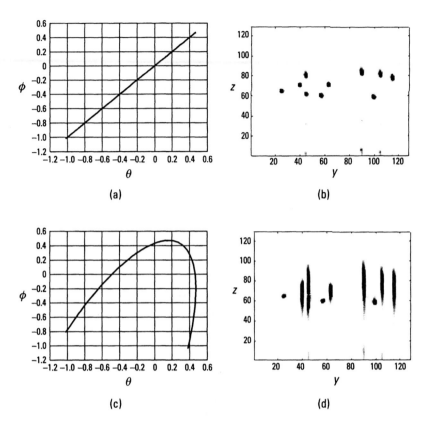

Figure 6.11 Point-scatterer simulation showing the effect of 3D motion on image formation: (a) assumed 2D motion; (b) resulting image after motion compensation; (c) assumed 3D motion; and (d) resulting image after motion compensation. (*Source:* [21].)

translation motion has already been removed from the data and the time-varying phase of a point-scatterer is given by

$$P_i(t) = y_i\theta(t) + z_i\phi(t) \tag{6.14}$$

If θ and ϕ are linearly related via $\theta(t) = a\phi(t)$, then the phases of any two point-scatterers

$$P_1(t) = (y_1 a + z_1)\phi(t) \tag{6.15}$$
$$P_2(t) = (y_2 a + z_2)\phi(t)$$

must also be linearly related as follows:

$$P_2(t) = \frac{(y_2 a + z_2)}{(y_1 a + z_1)} P_1(t) \qquad (6.16)$$

We conclude that if the rotational motion of the target is 2D, then the phase relationship between two scatterers must be related linearly. Conversely, if the phase relationship is not linear, there must exist 3D rotational motion on the target. Thus, by extracting the phases of two or more point-scatterers on the target using the AJTF algorithm and measuring the degree of nonlinearity between them, we can detect the presence of 3D motion.

We demonstrate the concept using a set of measured radar data of an aircraft. Figure 6.12 shows the degree of phase nonlinearity detected from 20 discrete frames in the data set. The corresponding imaging interval for each frame is 2.3 seconds, while the total flight duration is approximately 5 minutes. As we can see, there are several frames where 3D motions are significant. To corroborate this result, we examine the actual motion data at frames 2 and 18. Frame 2 has very small detected phase nonlinearity while frame 18 has very large phase nonlinearity. Shown in Figure 6.13(a, b) are the corresponding θ versus ϕ plots obtained for these two frames from the truth motion data. The actual motions are shown in the solid curves while the dashed lines are the best-fit linear approximations. It is clear from Figure 6.13(a) that the target motion in frame 2 is indeed very close to a pure 2D one. Figure 6.13(b) for frame 18, on the other hand, shows that the motion during this frame deviates significantly from a 2D one. Thus, the detected phase nonlinearity is a good indicator of 3D target motion. Figure 6.13(c, d) shows the respective images formed at these two frames using the AJTF motion compensation algorithm. As expected, the image

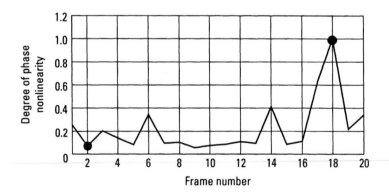

Figure 6.12 Detected phase nonlinearity indicating 3D motion from aircraft radar data. (*Source:* [21].)

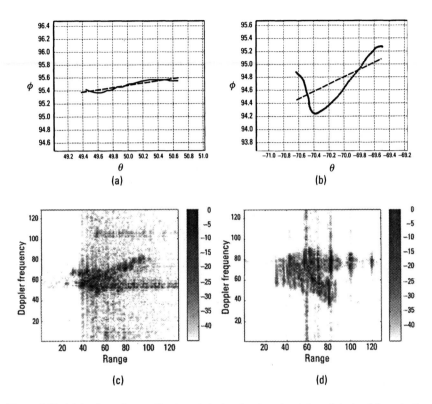

Figure 6.13 (a) Look angles on the target during the imaging interval derived from truth motion data for frame 2. The solid curve is the actual motion and the dashed line is the best-fit linear approximation; (b) look angles for frame 18 showing significant 3D motion; (c) resulting motion-compensated image for frame 2 showing a well-focused image; and (d) motion-compensated image for frame 18 showing significant blurring in the Doppler dimension.

for frame 2 is very well focused while the image for frame 18 is blurred in the Doppler dimension.

To summarize, we have shown that the existing motion compensation algorithm based on the 2D motion assumption cannot properly focus targets exhibiting 3D rotational motion during the imaging interval. We also showed that it is possible to detect the presence of 3D motion by measuring the phase nonlinearity between two or more point-scatterers on the target. The phase estimation can again be achieved effectively using the AJTF algorithm. Therefore, it is possible to distinguish the time intervals when the target undergoes smooth 2D motion from those containing more chaotic 3D motion. As a result, the good imaging intervals, where focused images are more easily formed, can be pinpointed in the data.

References

[1]   Chen, C. C., and H. C. Andrews, "Target Motion Induced Radar Imaging," *IEEE Trans. Aerospace Electronic Systems*, Vol. AES-16, January 1980, pp. 2–14.

[2]   Ausherman, D. A., et al., "Developments in Radar Imaging," *IEEE Trans. Aerospace Electronic Systems*, Vol. AES-20, July 1984, pp. 363–400.

[3]   Jain, A., and I. Patel, "SAR/ISAR Imaging of a Nonuniformly Rotating Target," *IEEE Trans. Aerospace Electronic Systems*, Vol. AES-28, January 1992, pp. 317–321.

[4]   Calloway, T. M., and G. W. Donohoe, "Subaperture Autofocus for Synthetic Aperture Radar," *IEEE Trans. Aerospace Electronic Systems*, Vol. AES-30, April 1994, pp. 617–621.

[5]   Itoh, T., H. Sueda, and Y. Watanabe, "Motion Compensation for ISAR Via Centroid Tracking," *IEEE Trans. Aerospace Electronic Systems*, Vol. AES-32, July 1996, pp. 1191–1197.

[6]   Wahl, D. E., et al., "Phase Gradient Autofocus—a Robust Tool for High Resolution SAR Phase Correction," *IEEE Trans. Aerospace Electronic Systems*, Vol. AES-30, July 1994, pp. 827–835.

[7]   Werness, S., et al., "Moving Target Imaging Algorithm for SAR Data," *IEEE Trans. Aerospace Electronic Systems*, Vol. AES-26, January 1990, pp. 57–67.

[8]   Carrara, W. G., R. S. Goodman, and R. M. Majewski, *Spotlight Synthetic Aperture Radar: Signal Processing Algorithms*, Norwood, MA: Artech House, 1995.

[9]   Chen, V. C., "Reconstruction of Inverse Synthetic Aperture Images Using Adaptive Time-Frequency Wavelet Transforms," *SPIE Proc. on Wavelet Applications*, Vol. 2491, April 1995, pp. 373–386.

[10]   Chen, V. C., and S. Qian, "Joint Time-Frequency Transform for Radar Range-Doppler Imaging," *IEEE Trans. Aerospace Electronic Systems*, Vol. AES-34, April 1998, pp. 486–499.

[11]   Qian, S., and D. Chen, "Decomposition of the Wigner-Ville Distribution and Time-Frequency Distribution Series," *IEEE Trans. Signal Processing*, Vol. SP-42, October 1994, pp. 2836–2842.

[12]   Ling, H., Y. Wang, and V. Chen, "ISAR Image Formation and Feature Extraction Using Adaptive Joint Time-Frequency Processing," *SPIE Proc. on Wavelet Applications*, Vol. 3708, April 1997, pp. 424–432.

[13]   Wang, Y., H. Ling, and V. C. Chen, "ISAR Motion Compensation Via Adaptive Joint Time-Frequency Technique," *IEEE Trans. Aerospace Electronic Systems*, Vol. AES-34, April 1998, pp. 670–677.

[14]   Qian, S., and D. Chen, "Signal Representation Using Adaptive Normalized Gaussian Functions," *Signal Processing*, Vol. 36, March 1994, pp. 1–11.

[15]   Qian, S., D. Chen, and Q. Yin, "Adaptive Chirplet Based Signal Approximation," *Proc. ICASSP*, Vol. III, May 1998, pp. 1871–1874.

[16]   Bultan, A., "A Four-Parameter Atomic Decomposition of Chirplets," *IEEE Trans. Signal Processing*, Vol. 47, March 1999, pp. 731–745.

[17]   Li, J., et al., "Comparison of High-Resolution ISAR Imageries from Measurement Data and Synthetic Signatures," *SPIE Proc. on Radar Processing, Technology, and Applications*, Vol. 3810, July 1999, pp. 170–179.

[18]   Chen, V. C., and W. J. Miceli, "Effect of Roll, Pitch and Yaw Motions on ISAR Imaging," *SPIE Proc. on Radar Processing, Technology, and Applications*, Vol. 3810, July 1999, pp. 149–158.

[19]   Rihaczek, A. W., and S. J. Hershkowitz, "Choosing Imaging Intervals for Identification of Small Ships," *SPIE Proc. on Radar Processing, Technology, and Applications*, Vol. 3810, July 1999, pp. 139–148.

[20]   Ling, H., and J. Li, "Application of Adaptive Joint Time-Frequency Processing to ISAR Image Formation," *Proc. of Tenth IEEE Workshop on Statistical Signal and Array Processing*, August 2000, pp. 476–479.

[21]   Li, J., H. Ling, and V. C. Chen, "An Algorithm to Detect the Presence of 3D Target Motion from ISAR Data," *Multidimensional Systems and Signal Processing*, Special Issue on Radar Signal Processing and Its Applications, 2001.

7

SAR Imaging of Moving Targets

SAR images are a high-resolution map of surface target areas and terrain in the range and the cross-range dimension [1–4]. If there are moving targets in the scene, SAR cannot simultaneously produce clear images of both these stationary targets and moving targets. Usually, moving targets appear as defocused and spatially displaced objects superimposed on the SAR map. Therefore, how to detect and clearly image moving targets becomes an important issue.

There are three basic issues for SAR imaging of moving targets: (1) how to detect moving targets [moving target indication (MTI)] in the background of stationary objects called clutter; (2) how to focus images of moving targets; and (3) how to place the detected moving targets into their true location in the SAR scene.

Using the MTI function, radar returns from terrain and stationary objects can be suppressed; only the returns from moving targets are used to reconstruct radar images. For focusing the image of detected moving targets, many algorithms that compensate for the target's motion and make phase corrections can be used. Because of the additional Doppler shift caused by target motion, the detected and focused target is not necessarily located in its true location in the SAR scene. To relocate it, a multiple-aperture antenna array may be used in the SAR system.

We will discuss radar returned signals from moving targets in Section 7.1 and analyze the effect of target motion on SAR image in Section 7.2. Then, we will review various approaches to detection and imaging of moving targets in Section 7.3. Finally, in Section 7.4 we will introduce how time-frequency transforms can be used for SAR imaging of moving targets.

7.1 Radar Returns of Moving Targets

When targets are moving, the motion-induced phase errors, which interact with the matched filter processing or cross-range compression, cause these images to be mislocated in the cross-range dimension and smeared in both the cross-range and the range domains.

Figure 7.1 illustrates the geometry of a side-looking SAR and a moving target. An aircraft carrying a radar platform is flying along the x-direction with a speed v and at an altitude h. The radar transmits a LFM signal

$$s_T(t) = \exp\left\{j2\pi\left(f_0 t + \frac{\eta}{2}t^2\right)\right\} \quad (0 \le t \le T) \tag{7.1}$$

where f_0 is the carrier frequency, η is the frequency modulation rate, T is the time duration of the signal, and the amplitude is normalized to unity.

The returned signal from a stationary point-scatterer is

$$s_R(t) = C_0 \exp\left\{j2\pi\left[f_0(t-\tau) + \frac{\eta}{2}(t-\tau)^2\right]\right\} \quad (0 \le t \le T) \tag{7.2}$$

where the constant C_0 is determined by the scatterer's reflectivity and the antenna's two-way azimuth pattern, $\tau = 2R/c$ is the two-way time delay,

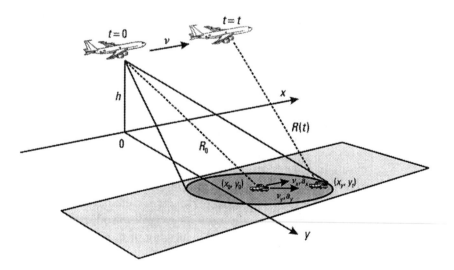

Figure 7.1 Geometry of an airborne side-looking SAR and a moving target.

and R is the range from the radar to the stationary point-scatterer. For simplicity, we assume that the reflectivity of the scatterer is 1. Thus, the baseband signal becomes

$$s_B(t) = C_0 \exp\left\{-j2\pi\left[f_0\tau - \frac{\eta}{2}(t-\tau)^2\right]\right\} \tag{7.3}$$

$$= C_0 \exp\{-j\Phi(t)\} \quad (0 \le t \le T)$$

where the phase function

$$\Phi(t) = 2\pi\left[f_0\tau - \frac{\eta}{2}(t-\tau)^2\right] = \frac{4\pi R}{\lambda} - \pi\eta\left(t - \frac{2R}{c}\right)^2 \tag{7.4}$$

and the wavelength $\lambda = c/f_0$, where c is the speed of electromagnetic wave propagation.

After range compression, the baseband output of the return from the stationary scatterer can be expressed as

$$s_0(t) = C_0 \exp\left\{-j\frac{4\pi R}{\lambda}\right\} \operatorname{sinc}\left[\pi\eta\delta T\left(t - \frac{2R}{c}\right)\right] \tag{7.5}$$

where δT is the width of the compressed pulse. When $t = 2R/c$, $\operatorname{sinc}[\pi\eta\delta T(t - 2R/c)] = 1$ and the peak value of the baseband output signal becomes

$$s_0(t = 2R/c) = C_0 \exp\left\{-j\frac{4\pi R}{\lambda}\right\} \tag{7.6}$$

7.1.1 Range Curvature

For a side-looking SAR, if the radar is moving along the x-direction with a speed of v, then the range from the radar to the point-scatterer becomes a function of x and can be expressed as follows [1]:

$$R(x) \cong R_{\min} + \frac{x^2}{2R_{\min}} \tag{7.7}$$

where $x = vt$ and $R_{\min}$ is the minimum range between the radar and the point-scatterer along the flight track. At time $t = 0$, the range reaches its minimum value of $R_{\min}$.

Assume during the time interval ΔT (i.e., from $t = -\Delta T/2$ to $t = +\Delta T/2$) that the radar can see the point-scatterer within its antenna beam-width. At time $t = \pm\Delta T/2$, the range reaches its maximum value $R_{\min} + \dfrac{(\Delta T/2)^2}{2R_{\min}}$. Thus, the range walk induced by the radar motion, which is also called the range curvature (shown in Figure 7.2), is

$$\Delta R = \frac{(\Delta T/2)^2}{2R_{\min}}$$

Therefore, to eliminate the range walk, we must carefully select the interval ΔT and the minimum range $R_{\min}$ to make sure that ΔR is within the size of range resolution cell.

7.1.2 Clutter Bandwidth

By substituting R in (7.6) with (7.7), the signal can also be expressed as a function of x:

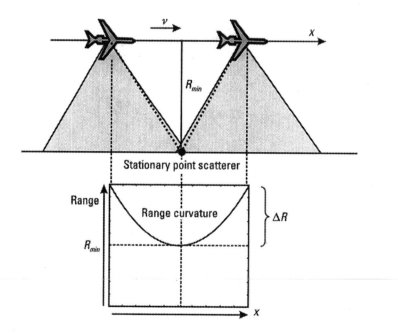

Figure 7.2 Range curvature induced by radar motion.

$$s_0(x) = C_0 \exp\left\{-j\frac{4\pi R(x)}{\lambda}\right\}$$

$$= C_0 \exp\left\{-j\frac{4\pi R_{min}}{\lambda}\right\} \exp\left\{-j\frac{2\pi}{\lambda}\frac{x^2}{R_{min}}\right\} \quad (7.8)$$

$$= C \exp\left\{-j\frac{2\pi}{\lambda}\frac{x^2}{R_{min}}\right\}$$

where C includes the constant C_0 and a constant phase $4\pi R_{min}/\lambda$. The equation (7.8) is the representation of the stationary scatterer in the cross-range domain. The phase of the output base-band signal of the stationary scatterer becomes

$$\Phi(x) = \frac{2\pi}{\lambda}\frac{x^2}{R_{min}} \quad (7.9)$$

After applying a matched filtering to the base-band signal in (7.8), the quadratic phase (7.9) is removed and a clear image of the stationary point-scatterer can be seen.

For detection and imaging of moving targets, stationary objects are usually considered as clutter. If we know the bandwidth of the clutter, it may be suppressed by filtering in the frequency domain.

The clutter bandwidth is equal to the Doppler bandwidth of the base-band signal returned from stationary objects, which can be derived from the time derivative of the phase function (7.9)

$$BW_{Clutter} = \frac{1}{2\pi}\frac{d\Phi(x = vt)}{dt} = \frac{2}{\lambda}\frac{vt}{R_{min}}v \quad (7.10)$$

The magnitude of the maximum clutterer bandwidth is

$$|BW_{Clutter}| = \frac{2}{\lambda}\frac{v\Delta T}{R_{min}}v \cong \frac{2}{\lambda}\beta_{3dB}v \quad (7.11)$$

where

$$\beta_{3dB} \cong \frac{v\Delta T}{R_{min}}$$

is the 3-dB antenna beamwidth at range R_{min} without coherent Doppler processing and was derived in [5]. Knowing the maximum clutter bandwidth, clutter can be suppressed by applying a high-pass filtering to suppress energy within the clutter bandwidth.

7.1.3   Analysis of Radar Returns from Moving Targets

Based upon the discussion on radar returns from stationary targets, we can now discuss radar returns from moving targets. Assume that at $t = 0$ a point target is located at (x_0, y_0) and the radar platform is located at $(x = 0, y = 0, z = h)$ as illustrated in Figure 7.1. When the target moves with a velocity v_y and acceleration a_y in the radial direction, and a velocity v_x and acceleration a_x in the x-direction, and when the radar moves with a velocity v along the x-direction, then at time t the target moves to $(x_0 + v_x t + a_x t^2/2, y_0 + v_y t + a_y t^2/2, 0)$ and the radar moves to $(x = vt, y = 0, z = h)$ as shown in the figure. The range from the radar to the point target at time t becomes

$$R(t) = [(vt - v_x t - a_x t^2/2 - x_0)^2 + (y_0 + v_y t + a_y t^2/2)^2 + h^2]^{1/2}$$
$$= [R^2_{radar}(t) + R^2_{target}(t)]^{1/2} \qquad (7.12)$$

where the range component due to the radar motion is

$$R_{radar}(t) = [(vt - x_0)^2 + y_0^2 + h^2]^{1/2}$$

and the range component due to the target motion is

$$R_{target}(t) = [(v_x t + a_x t^2/2)^2 + (v_y t + a_y t^2/2)^2$$
$$- 2(vt - x_0)(v_x t + a_x t^2/2) + 2y_0(v_y t + a_y t^2/2)]^{1/2}$$

By taking Maclaurin series expansion

$$R(t) \cong R(t)|_{t=0} + \frac{dR(t)}{dt}\bigg|_{t=0} t + \frac{1}{2}\frac{d^2 R(t)}{dt^2}\bigg|_{t=0} t^2 \qquad (7.13)$$

a simplified expression of $R(t)$ can be derived as

$$R(t) \cong R_0 + \frac{x_0 v_x + y_0 v_y - x_0 v}{R_0} t + \frac{v^2 + v_x^2 + v_y^2 + x_0 a_x + y_0 a_y - 2vv_x}{2R_0} t^2$$

$$(7.14)$$

where $R_0 = R(t)\big|_{t=0} = (x_0^2 + y_0^2 + h^2)^{1/2} \gg x_0$ is the initial range at $t = 0$.

Because x_0 is much smaller than R_0 and y_0 is approximately equal to R_0, then the baseband return from the moving target can be derived from (7.6) as

$$s_0(t) = C_0 \exp\left\{-j4\pi \frac{R(t)}{\lambda}\right\}$$

$$\cong C \exp\left\{-j2\pi \frac{2v_y}{\lambda} t\right\} \exp\left\{-j\frac{4\pi}{\lambda}[(v - v_x)^2 + v_y^2 + R_0 a_y]\frac{t^2}{2R_0}\right\}$$

$$= C \exp\{-j\Phi_{Shift}(t)\} \exp\{-j\Phi_{Defocus}(t)\} \qquad (7.15)$$

where

$$\Phi_{Shift}(t) = 2\pi \frac{2v_y}{\lambda} t \qquad (7.16)$$

is a linear phase function due to the target velocity in the radial direction, and

$$\Phi_{Defocus}(t) = \frac{4\pi}{\lambda}[(v - v_x)^2 + v_y^2 + R_0 a_y]\frac{t^2}{2R_0} \qquad (7.17)$$

is a quadratic phase function determined by the relative velocity between the radar and the moving target in the x-direction, the radial velocity, and acceleration of the target. From the above phase functions we can see that by the use of a matched filter designed to match the baseband returns from a stationary target, the linear phase change due to the target's radial velocity v_y in (7.16) causes the image of the moving target to be shifted in the cross-range direction, and the quadratic phase variation in (7.17) causes the image of the moving target to be defocused as illustrated in Figure 7.3.

Now we can examine how the Doppler frequency shift is affected by radar and target motions. Because time-derivative of the phase function $2\pi[2R(t)/\lambda]$ in (7.6) is the Doppler frequency shift f_D, we have

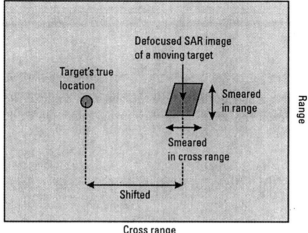

Figure 7.3 Effect of target motion on SAR imaging.

$$f_D = \frac{2}{\lambda}\frac{dR(t)}{dt} \tag{7.18}$$

where the range $R(t)$ is given by (7.14). Therefore, the Doppler shift consists of two parts:

$$f_D = f_{D_{Radar}} + f_{D_{Target}}$$

where the Doppler shift due to the radar motion is

$$f_{D_{Radar}} = -\frac{2}{\lambda}\frac{x_0 v}{R_0} + \frac{2}{\lambda}\frac{v^2}{R_0}t \tag{7.19}$$

and the Doppler shift due to the target motion is

$$f_{D_{Target}} = -\frac{2}{\lambda}\frac{x_0 v_x + y_0 v_y}{R_0} + \frac{2}{\lambda}\frac{v_x^2 + v_y^2 + x_0 a_x + y_0 a_y - 2vv_x}{R_0}t \tag{7.20}$$

From (7.19) and (7.20), we can analyze the effects of the target motion on SAR imaging.

7.2 The Effect of Target Motion on SAR Imaging

For a stationary target, according to (7.19), the Doppler shift is induced only by the radar motion, and consists of two parts: the Doppler centroid

$$f_{D_{RadarC}} = -\frac{2}{\lambda}\frac{x_0 v}{R_0} \tag{7.21}$$

and the Doppler rate

$$f_{D_{RadarR}} = \frac{2}{\lambda}\frac{v^2}{R_0}t \tag{7.22}$$

Given a radar velocity v and an initial range R_0, the Doppler centroid of the stationary target is determined only by its location x_0 in the cross-range domain as shown in (7.21). After the matched filter processing, the Doppler rate in (7.22) can be removed, and the Doppler shift becomes a constant Doppler centroid. Thus, by taking the Fourier transform along the cross-range domain, the image of the stationary target is clearly focused.

However, for a moving target, according to (7.20) its Doppler centroid is determined not only by its geometric location (x_0, y_0), but also by its velocity (v_x, v_y). The Doppler drift of the moving target becomes

$$f_{D_{TargetC}} = -\frac{2}{\lambda}\frac{x_0 v_x + y_0 v_y}{R_0} \tag{7.23}$$

and its Doppler rate becomes

$$f_{D_{TargetR}} = \frac{2}{\lambda}\frac{v_x^2 + v_y^2 + x_0 a_x + y_0 a_y - 2 v v_x}{R_0}t \tag{7.24}$$

From (7.23) and (7.24) we know that if the target moves only in the radial direction (i.e., $v_x = 0$; $v_y \neq 0$), then the image shift in the cross-range direction is determined only by the target's radial velocity v_y. If the target has only motion in the x-direction (i.e., $v_x \neq 0$; $v_y = 0$), then the image shift in the cross-range direction, which is determined by a small value $(x_0/R_0)v_x$, can be negligible. However, in both cases the image of the moving target is defocused in the cross-range domain. A target's movement through range cells, called range walk, can also cause the image to be smeared in the range domain.

As we mentioned earlier, while stationary targets are well focused, the image of moving targets may become defocused and shifted in the cross-range domain. Figure 7.4 illustrates a SAR image of a moving point target (target 2) compared with a SAR image of a stationary point target (target 1). Figure 7.4(a) is the SAR image with the two targets. The one near the center is the stationary target and the smeared one on the left is the moving target. Figure 7.4(b) shows the time-Doppler frequency distribution of the two targets, where the time-scale is normalized to 1 and the frequency scale is normalized to 0.5. The horizontal line of the time-Doppler frequency distribution is from the stationary target, and the slope ramp of the time-

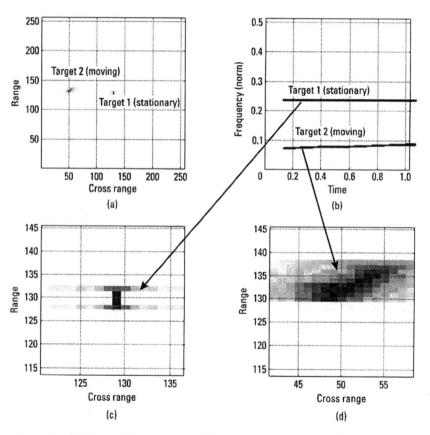

Figure 7.4 SAR image of a moving target 2 compared with SAR image of a stationary target 1: (a) SAR image of the two targets; (b) time-Doppler distributions of the two targets; (c) close-up of the image of the stationary target 1; and (d) close-up image of the moving target 2—image is defocused, smeared, and shifted from its true location.

Doppler frequency distribution is from the moving target. Since image focusing is based on the stationary target [Figure 7.4(c)], the moving target is defocused, smeared, and shifted from its true location [Figure 7.4(d)], where Figure 7.4(c, d) are close-ups of the two targets in Figure 7.4(a). In Figure 7.4(d), we can see that the defocusing in the cross-range direction is due to the focus processing based on the stationary target, the smearing in the range domain is caused by the target's range walk, and the shift in the cross-range direction is due to the target's radial motion.

If we can focus on the moving target by removing its Doppler rate $f_{D_{TargetR}}$, then the image of stationary targets becomes defocused. We can only focus on either the stationary target or the moving target but not on both simultaneously.

7.3 Detection and Imaging of Moving Targets

Raney first discussed the issue of SAR imaging of moving targets in [6]. He proposed a simple frequency-domain technique for detecting moving targets in a single-aperture channel SAR. Since the detection requires clutter suppression, clutter cancellation approaches using multiple antenna channels, such as the space-time adaptive processing (STAP), can also be used. In this section, we will review various approaches using single channel and multiple channels to SAR imaging of moving targets.

7.3.1 Single-Aperture Antenna SAR

As we described earlier in Sections 7.1 and 7.2, to generate a focused image of a moving target, accurate estimate of its phase history function, which is determined by the Doppler centroid and the Doppler rate of the moving target, is necessary. Any error in estimating the Doppler rate can significantly defocus the image of the moving target. Though the error in estimating the Doppler centroid does not affect image focusing, it can increase the ambiguity of the target location in the cross-range. The following approaches, which estimate the Doppler centroid and the Doppler rate for each individual moving target, have been proposed [7].

7.3.1.1 Subaperture Focusing

In the synthetic aperture processing, since longer aperture time causes larger motion-induced phase errors and, thus, image degradation, a method called subaperture processing can be used to compensate a target's motion [8]. In the subaperture processing, the full aperture time is broken into a sequence of short aperture time or subaperture, and for each subaperture a low-resolution, but well-focused image, is formed. Within each subaperture, if the radar signal can be characterized by a linear FM, the differences of the motion-induced Doppler shifts between adjacent subaperture images can be determined by using correlation technique and are corrected in a piecewise manner. Thus, by coherently combining the corrected subaperture images a high-resolution image can be generated as illustrated in Figure 7.5.

The subaperture method is equivalent to a time gating approach. It can produce focused images of moving targets, improve the signal-to-clutter ratio, and suppress spurious artifacts.

7.3.1.2 Filter-Bank Approach

Based on the fact that the Doppler frequency shift of a moving target is different from that of stationary objects, Raney proposed a filter-bank approach for the detection and estimation of moving targets in the frequency domain [6]. Because the clutter occupies a certain frequency bandwidth, in order to detect a moving target the Doppler shift of the moving target must be greater than the clutter bandwidth given in (7.11):

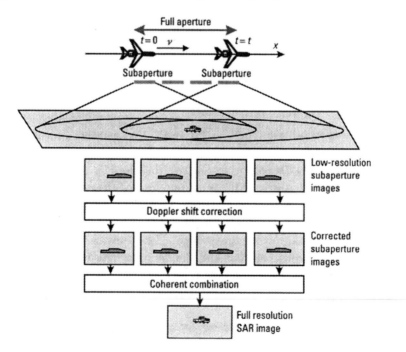

Figure 7.5 Subaperture processing for image focusing.

$$|f_{D_{TargetC}}| > |BW_{Clutter}|$$

From (7.11) and (7.23), we have

$$v_y > v\beta_{3dB}$$

where we assume $R_0 >> x_0$ and $R_0 \cong y_0$.

A minimum detectable velocity can be expressed as $(v_y)_{min} = Kv\beta_{3dB}$, which depends on the velocity of the radar platform, the antenna beamwidth, and a constant K selected by the desired false-alarm and detection rates. If the radial velocity of a moving target is less than the minimum detectable velocity $(v_y)_{min}$, the target's Doppler frequency will be buried in the clutter bandwidth and cannot be detected. We also notice that the minimum detectable velocity does not depend on the range R_0.

To improve signal-to-clutter ratio, the radar PRF should be much greater than the clutter bandwidth. Thus, there will be a large frequency region beyond the clutter bandwidth which moving targets may fall in.

To detect moving targets in the frequency domain, a simple approach is to prefilter the data before performing the cross-range compression. A bank of Doppler filters is usually used as the prefilter that covers the entire Doppler frequency region.

Freeman and Currie proposed an approach for further improvement of the Doppler filter-bank method [5]. The approach makes the moving target signals alias onto zero Doppler frequency by using down-sampling method. Therefore, the conventional matched filter that has a reference function for stationary targets and is centered at the zero Doppler frequency can still be used for the cross-range compression.

Because radar signal is transmitted at a certain PRF, Doppler ambiguities (i.e., the observed Doppler frequency of the moving target can be due to two or more frequency shifts), may occur and the clutter-band repeats at intervals of the PRF. Thus, the radial velocity is also ambiguous, and the blind speed intervals are raised as indicated in [5]:

$$v_y = \frac{\lambda}{2}(f_{D_{TargetC}} + nPRF) \tag{7.25}$$

where n is an integer. This makes it more difficult to estimate the target's radial velocity and to distinguish moving targets from stationary targets.

The Doppler rate $f_{D_{TargetR}}$ of a moving target is usually needed for focusing the image of the moving target. Similar to the Doppler filter-bank, a Doppler-rate filter-bank may be used to compensate the unknown Doppler-rate of the moving target before applying the reference signal to it.

To estimate the Doppler rate, an algorithm based on the Doppler centroid was proposed in [5]. First, the cross energy between the received baseband signal $s_0(t)$ and a reference signal $s_{Ref}(t, \eta) = \exp\{j2\pi(\eta/2)t^2\}$ is calculated:

$$|\epsilon(\eta)| = \frac{1}{T}\left| \int\limits_{-T/2}^{T/2} s_0(t, f_{D_R})s_{Rref}(t, \eta)dt \right| \qquad (7.26)$$

where η is the Doppler rate to be estimated. After removing the Doppler shift in (7.15), the received baseband signal can be written as $s_0(t, f_{DR}) = \exp\{j2\pi(f_{D_R}/2)t^2\}$, where

$$f_{D_R} = \frac{2}{\lambda R_0}[(v - v_x)^2 + v_y^2 + R_0 a_y]$$

and, thus, we have

$$|\epsilon(\eta)| = \left| K \int\limits_{-T/2}^{T/2} \exp\{j\pi(f_{D_R} - \eta)^2\}dt \right| \qquad (7.27)$$

If $f_{D_R} - \eta = 0$, the $|\epsilon(\eta)|$ has a maximum cross-energy. Thus, $\bar{\eta} = \max_{\eta}|\epsilon(\eta)|$ will be the estimated Doppler rate. However, to accurately estimate motion parameters requires a large number of filters.

7.3.1.3 Parameter Estimation for Doppler and Doppler-Rate

The filter-bank approach matches the Doppler and the Doppler-rate separately. If the Doppler and the Doppler-rate can be estimated simultaneously, a better result of target detection and imaging can be expected.

A parameter estimation approach based on the maximum likelihood method was proposed in [9]. Estimation of motion parameters based on a sequence of single-look SAR images was described in [10, 11]. A method called keystone formatting that uses a unique processing kernel of 1D interpolation of deramped phase history was proposed for SAR imaging of moving targets in [12].

7.3.2   Multiple-Antenna SAR

Raney discussed the effect of motion parameters on the image of a moving target (i.e., the problem of image smearing and mislocating [6]). For a special case where $x_0 = 0$ and $a_x = a_y = 0$, we have

$$f_{D_{RadarC}} = 0 \tag{7.28}$$

$$f_{D_{RadarR}} = -\frac{2}{\lambda} \frac{v^2}{R_0} t \tag{7.29}$$

$$f_{D_{TargetC}} \approx -\frac{2}{\lambda} \frac{y_0 v_y}{R_0} \cong -\frac{2}{\lambda} v_y \tag{7.30}$$

and

$$f_{D_{TargetR}} = -\frac{2}{\lambda} \frac{v_x^2 + v_y^2 - 2vv_x}{R_0} t \tag{7.31}$$

If we know the Doppler centroid and the Doppler rate exactly, then the velocity (v_x, v_y) of the moving target can be calculated from the above equations.

However, when $x_0 \neq 0$, the Doppler centroid and the Doppler rate of the moving target are determined not only by its velocity (v_x, v_y), but also by its initial location (x_0, y_0), which we may not know. Therefore, the velocity of a moving target cannot be obtained, and the mislocating problem cannot be solved.

To estimate the target's velocities and replace mislocated moving targets to their true locations, a multiple-antenna system, such as the interferometry, planar apertures, and antenna array, is required. Having multiple antennas with their independent receive channels, the so-called displaced phase center antenna (DPCA) technique or the STAP technique can be applied to suppress clutter.

DPCA was first proposed by F. R. Dickey and M. M. Santa in 1953 in a technical report of General Electric [13]. DPCA technique was motivated by the two-pulse cancellation technique in MTI. Two side-looking antenna apertures are along the flight track and normal to the radar LOS. The PRF of the transmitted signals is adjusted such that if a pulse is transmitted at the first aperture, then the second aperture will transmit a pulse when it moves to the position where the first aperture was located and transmitted the previous pulse. With the DPCA, targets that are buried in the clutter and cannot be detected using single channel methods may be detectable [14, 15].

The DPCA uses only two antenna-apertures. If multiple apertures are used, the radar receives a set of returns, each stamped by its time of arrival as well as by its spatial location at the apertures, and the processing is referred to as a space-time processing. The STAP proposed in the 1970s is a processing that has adaptive spatial and temporal weights [16, 17]. Later, Klemm investigated variations to the ideal STAP for airborne MTI [18, 19].

With multiple-antenna and STAP, SAR is able to detect moving targets and produce a range and cross-range image of both stationary targets and moving targets.

7.3.2.1   Ground Moving Target Indicator

Ground moving target indicator (GMTI) is designed to reject radar returns from clutter, such as buildings and trees, and detect moving targets, such as tanks, trucks, and aircraft, that could otherwise be obscured. The difference of Doppler frequency shifts between moving targets and the clutter is used to suppress the clutter and detect moving targets. With the GMTI, slow moving targets can also be detected even if their Doppler shifts are very small.

There are several multiple-antenna techniques to perform GMTI. The basic idea behind using multiple antennas is to have multiple phase centers. The radar compares the received data at the same place in space but at different times. When the radar platform moves, the phase center of each antenna passes through the same place at different times; this is called displaced phase center. With displaced phase centers, the clutter will remain the same, but moving targets will change their locations [20].

In the GMTI radar system, instead of a single antenna aperture, an antenna array is used. The antenna array may be either along the flight track such as the Joint Strategic Target Attack Radar System (Joint STARS) [21, 22], which is an electronically-scanned side-look airborne radar, or a scanned planar array mounted on the nose of an airplane such as the AN/APY-6 radar [23, 24]. Both the Joint STARS and the AN/APY-6 are three-port interferometric radars. Compared with the conventional mono-pulse radar, the interferometric radar provides a low sidelobe radiation pattern.

The AN/APY-6 radar is a new generation of the precision surveillance, tracking, and targeting radar operating at X-band and designed for use in the littoral area. This kind of radar, including the Ku-band Gray Wolf AN/APG-76 radar [25], provides the utility of detecting and locating both stationary and moving targets. The antenna of the AN/APY-6 radar is a mechanically scanned four-aperture planar array antenna. One larger aperture is used for transmitting waveforms for SAR and GMTI as well as for receiving radar returns for SAR. The three smaller apertures used for GMTI receive only channels [23].

As we discussed in Section 7.2, the image of a moving target may be shifted in the cross-range direction. By estimating the amount of shift, the true location of the target can be recovered. Two-port interferometry is able to estimate the target azimuth position and relocate the target to its true location. Yadin [25] indicated that the target angular location could be estimated by the differential phase of two interferomeric channels, $\frac{\lambda}{2\pi D} phase\{I_R, I_L\}$, where D is the spacing between the two receiving antennas, and I_R and I_L are the received signals of the right-channel and the left-channel, respectively. Actually, the differential phase in the two receivers can be derived as follows [26]:

$$\Delta\Phi \cong \frac{2\pi D}{\lambda} \frac{\Delta_{cr}}{R_0} \Delta_{Shift} \qquad (7.32)$$

where Δ_{cr} is the cross-range resolution of the image and Δ_{Shift} is the amount of shift in the cross-range of the image. Thus, if we can measure the differential phase $\Delta\Phi$, the amount of the shift Δ_{Shift} can be estimated by

$$\Delta_{Shift} \cong \frac{\lambda}{2\pi D} \frac{R_0}{\Delta_{cr}} \Delta\Phi \qquad (7.33)$$

The AN/APY-6 and APG-76 use three-port STAP clutter suppression interferometry to simultaneously cancel clutter and determine the true cross-range location of the detected target. The true location of the detected target is calculated using the interferometric phase, which is the phase difference between the phase residue of the left interferometric channel and that of the right interferometric channel. The interferometric phase is linearly proportional to the cross-range location of the target. By comparing the estimated azimuth location to the Doppler of the target, the offset of the Doppler can be measured. Then, the offset is used to correct the cross-range location of the target as described in [27]. After detecting and locating the target, the image of the target can also be focused.

Instead of using phase estimation, another approach to the estimation of the shift in cross-range is to apply the Fourier transform as described in Section 7.3.2.2.

7.3.2.2 Velocity SAR

Another multiple-antenna approach uses a relatively larger number of antennas along the direction of the flight track, such as the velocity SAR (VSAR) proposed by Friedlander and Porat [28] for the purpose of detecting and relocating moving targets. Having a single transmitting antenna and an array of receiving antennas, VSAR can produce a full 3D range, cross-range, and velocity complex SAR image as illustrated in Figure 7.6. The VSAR system has the capability of estimating the velocities for the scatterers in every range and cross-range cell. By separating scatterers in terms of their velocities, it is able to separate moving targets from clutter.

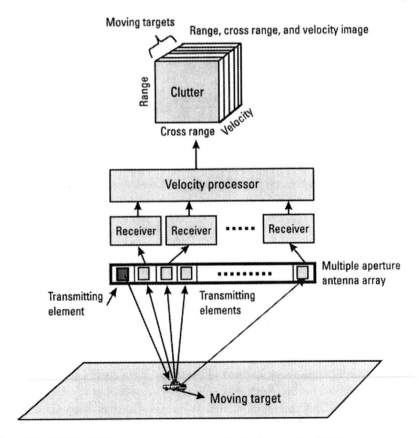

Figure 7.6 VSAR and 3D-range, cross-range, and velocity complex SAR image.

The VSAR processing is different from other multiple-antenna SAR. Without having a front-end STAP for clutter suppression, the VSAR produces complex SAR images for different velocities using the Fourier transform. With the velocity discrimination capability, a high degree of clutter rejection can be achieved.

With multiple phase centers in a linear antenna array, clutter can be suppressed. However, there are only two parameters involved (i.e., the number of receiving antennas and the spacing between two adjacent antennas). The two parameters physically limit the capability of estimating the accurate locations of fast moving targets, and only slow moving targets can be relocated. The multiple antenna approach that transmits signals with only a single carrier-frequency can be generalized to transmitting signals with multiple carrier-frequencies. With a multifrequency antenna array, the SAR not only can suppress the clutter, but can also detect and relocate both the slow and the fast moving targets as described in [26].

7.4 SAR Imaging of Moving Targets Using Time-Frequency Transforms

A new approach to SAR imaging of moving targets is based on the time-frequency transform. As we described in Section 7.3.1.1, the subaperture approach is actually a time-domain gating. It does not work well when energies of returned signals are overlapped in the time-domain. The filter-bank approach is a frequency filtering. It may fail when energies of returned signals are overlapped in the frequency domain. Time-frequency method takes the advantages of the subaperture approach and the filter-bank approach and combines the time and the frequency domains to form a new approach.

Figure 7.7(a) is a simulated SAR image with a stationary point target located around the center of the image. If a sequence of data is taken from received signals at range cells where the stationary target is located, the time-frequency distribution of the data sequence is shown in Figure 7.7(b), where the strong horizontal line corresponds to the stationary point target and other weaker distributions correspond to the clutter located at the same range cells. Figure 7.7(c) is a simulated SAR image with a moving point target located around the center of the image. The time-frequency distribution of a data sequence taken at range cells where the moving point target is located is shown in Figure 7.7(d), where a strong slope ramp corresponds to the moving point target and other weaker distributions correspond to the clutter located at the same range cells.

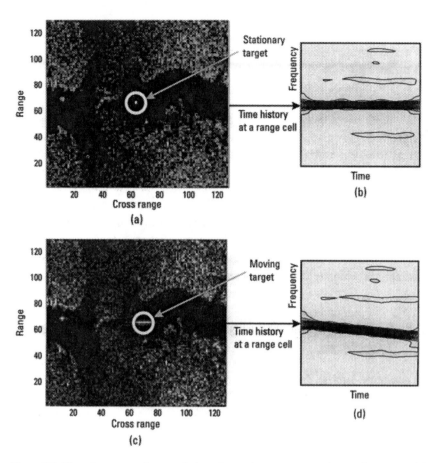

Figure 7.7 Time-frequency characteristics of a stationary target and a moving target: (a) a simulated SAR image with a stationary target; (b) time-frequency distribution of the stationary target; (c) a simulated SAR image with a moving target; and (d) time-frequency distribution of the moving target.

Because the time-frequency distribution of a moving target is always a slope ramp as shown in Figure 7.7(d), time-frequency transforms can be used to distinguish moving targets from the stationary ones.

7.4.1   Estimation of Doppler Parameters Using Time-Frequency Transforms

A target's motion information is usually embedded in phase shifts of radar returns. Time-frequency transform is a useful tool to estimate instantaneous phase shifts of scatterers on the target, and the phase-shift histories can then be used to produce a focused image of moving targets [29].

As illustrated in Figure 7.8, the basic procedure for the estimation of Doppler parameters is the following:

1. Calculate the time-frequency distribution of a data sequence taken from the received signals and at the range cells where the moving target is detected.

2. Extract ridges of the calculated time-frequency distribution.

3. Find a curve function that represents the strong ridge by using curve-fitting.

4. Take integration of the ridge curve function to generate the corresponding phase function.

5. Use the conjugate of the phase function to compensate the phase history of the unfocused target.

6. Take the Fourier transform to form a focused image of the moving target.

We noticed that because the focus processing is based on the Doppler rate (i.e., the slope of the time-frequency ridge curve), it is not affected by the mean value of the Doppler shift, which affects only the relative position of the moving target with respect to stationary background.

The time-frequency method can compensate for the relative translational motion between the radar and the target and works well for point-like targets (i.e., target dimension is smaller than the range resolution), or

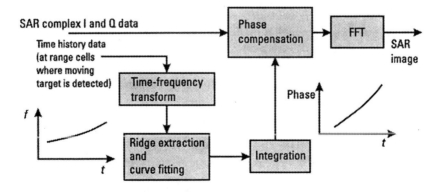

Figure 7.8 Estimate of Doppler rate and image focusing using the time-frequency transform.

for extended targets characterized by a dominant scatterer or by a set of scatterers which have similar reflectivity characteristics. In general, rotational motions of extended targets cannot be compensated very well. They must be compensated by other methods, such as the multiple PPP method described in Chapter 6 [2].

Usually, time-frequency distribution of the radar return from a strong scatterer, such as a corner reflector, can be used to accurately calculate the Doppler rate of the scatterer. However, when radar return is not strong enough, the time-frequency distribution cannot be used directly to estimate the Doppler rate. In this case, in addition to the time-frequency transform, Hough transform may be applied to find the slope of the time-frequency distribution [30, 31]. A combined space-time and time-frequency processing can also be used to improve the signal-to-disturbance ratio and estimate the instantaneous frequency of the moving target [32].

The Hough transform has been widely used for the detection of straight lines, parabolic geometry, ellipses, and other various curves in a 2D space [33]. It works well in a noisy environment and other complicated backgrounds. Duda and Hart [34] improved the Hough transform by using the angle-radius parameters:

$$r = x \cos \theta + y \sin \theta \qquad (7.34)$$

to replace the slope and intercept parameters, where (x, y) are the coordinates of a point on the curve, r indicates the perpendicular distance of the line from the center of the 2D space, and θ is the angle between the perpendicular to the line and the x-axis. For an $N \times N$ 2D space, the range of r is $-N\sqrt{2}$ to $+N\sqrt{2}$ and the range of θ is 0 to 2π. The Hough transform calculates all of the r values for every angle θ to determine whether the (r, θ) pair belongs to a curve. If the 2D space contains only straight lines, the Hough transform shows sharp peaks at the coordinates of (r, θ) where the lines lie. Figure 7.9(a) shows two straight lines in a 2D space, and Figure 7.9(b) is the Hough transform of the structures in the 2D space where the two peaks are the two detected straight lines.

7.4.2   Time-Frequency-Based SAR Image Formation for Detection of Moving Targets

The same principle of time-frequency-based image formation for ISAR [35, 36], as described in Chapter 5, can also be applied to SAR image formation. By the use of the time-frequency-based image formation, the

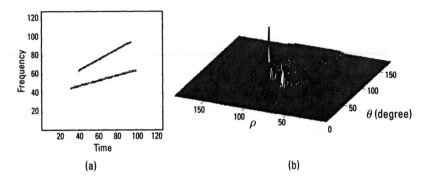

Figure 7.9 (a) Two straight lines in the time-frequency domain; and (b) the Hough transform of the two lines in the time-frequency domain, where the two peaks correspond to the two lines.

image of a moving target, which is smeared by using the Fourier transform, then becomes a sequence of clear images in the temporal image frame.

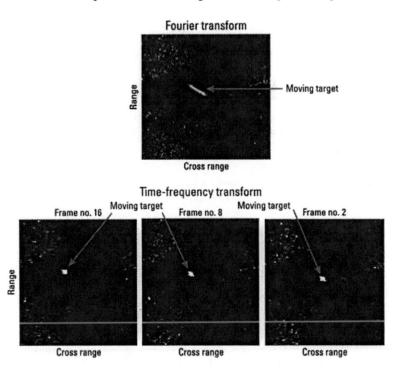

Figure 7.10 (a) The smeared SAR image generated by conventional Fourier-based image formation; and (b) three temporal image frames of the SAR image produced by the time-frequency-based image formation showing target motion.

As discussed in Chapter 5, the time-frequency-based image formation does not apply any phase compensation. Instead, it takes just time sampling to display the instantaneous Doppler shift for each scatterer center on the target. Figure 7.10 shows three temporal image frames of a SAR image produced by the time-frequency-based image formation. Without using conventional focusing methods, each temporal image frame is automatically focused for both the moving and the stationary scene. Just like the time-frequency-based ISAR image formation, the information contained in the multiple time-varying image frames can be used to detect moving targets and to focus images of moving targets.

References

[1]   Curlander, J. C., and R. N. McDonough, *Synthetic Aperture Radar—System and Signal Processing*, New York: John Wiley & Sons, 1991.

[2]   Carrara, W. G., R. S. Goodman, and R. M. Majewski, *Spotlight Synthetic Aperture Radar—Signal Processing Algorithms*, Norwood, MA: Artech House, 1995.

[3]   Wehner, D., *High-Resolution Radar, Second Edition*, Norwood, MA: Artech House, 1995.

[4]   Soumekh, M., *Synthetic Aperture Radar Signal Processing with MATLAB Algorithms*, New York: Wiley, 1999.

[5]   Freeman, A., and A. Currie, "Synthetic Aperture Radar (SAR) Images of Moving Targets," *GEC Journal of Research*, Vol. 5, No. 2, 1987, pp. 106–115.

[6]   Raney, R. K., "Synthetic Aperture Imaging Radar and Moving Targets," *IEEE Trans. on Aerospace and Electronic Systems*, Vol. AES-7, No. 3, 1971, pp. 499–505.

[7]   Li, F. K., et al., "Doppler Parameter Estimation for Spaceborne Synthetic-Aperture Radar," *IEEE Trans. on Geoscience and Remote Sensing*, Vol. 23, No. 1, 1985, pp. 47–56.

[8]   Wu, K. H., and M. R. Vant, "A SAR Focusing Technique for Imaging Targets with Random Motion," *NAECON*, 1984, pp. 289–295.

[9]   Barbarossa, S., "Detection and Imaging of Moving Targets with Synthetic Aperture Radar. Part 1: Optimal Detection and Parameter Estimation Theory," *IEE Proc.-F*, Vol. 139, No. 1, 1992, pp. 79–88.

[10] Kirscht, M., "Detection and Velocity Estimation of Moving Objects in a Sequence of Single-Look SAR Images," *Proc. IGARSS*, Lincoln, NE, Vol. 1, 1996, pp. 333–335.

[11] Kirscht, M., "Detection, Velocity Estimation and Imaging of Moving Targets with Single-Channel SAR," *Proc. EUSAR*, Germany, 1998, pp. 587–590.

[12] Perry, R. P., R. C. DiPietro, and R. L. Fante, "SAR Imaging of Moving Targets," *IEEE Trans. on Aerospace and Electronic Systems*, Vol. 35, No. 1, 1999, pp. 188–199.

[13]   Dickey, F. R., and M. M. Santa, "Final Report on Anti-Clutter Techniques," *General Electric Co., Heavy Military Electron. Dept.*, Rep. No. R65EMH37, Syracuse, NY, 1953.

[14]   Stone, M. L., and W. J. Ince, "Air-to-Ground MTI Radar Using a Displaced Phase Center Phased Array," *IEEE International Radar Conference*, 1980, pp. 225–230.

[15]   Sullivan, R. J., *Microwave Radar: Imaging and Advanced Concepts*, Norwood, MA: Artech House, 2000.

[16]   Brennan, L. E., and I. S. Reed, "Theory of Adaptive Radar," *IEEE Trans. AES*, Vol. 9, No. 2, 1973, pp. 237–252.

[17]   Brennan, L. E., J. D. Mallett, and I. S. Reed, "Adaptive Array in Airborne MTI Radar," *IEEE Trans. AP*, Vol. 24, No. 5, 1976, pp. 607–615.

[18]   Klemm, R., "Adaptive Airborne MTI: An Auxiliary Channel Approach," *IEE Proc-F.*, Vol. 134, No. 3, 1987.

[19]   Klemm, R., and J. Ender, "New Aspects of Airborne MTI," *Proc. IEEE International Radar Conference*, 1990, pp. 335–340.

[20]   Petterson, M. I., L. M. H. Ulander, and H. Hellsten, "Simulations of Ground Moving Target Indication in an Ultra-Wideband and Wide-Beam SAR System," *SPIE Proc. on Radar Processing, Technology, and Applications*, 1999, pp. 84–95.

[21]   Shnitkin, H., "A Unique Joint STARS Phased-Array Antenna" *Microwave Journal*, January 1991, pp. 131–141.

[22]   Shnitkin, H., "Joint Stars Phased Array Radar Antenna," *IEEE AES System Magazine*, October 1994, pp. 34–41.

[23]   Gross, L. A., R. A. Guarino, and H. Holt, "AN/APY-6 Real Time Surveillance and Targeting Radar Development," *NATO RTO Meeting Proc. on High Resolution Radar Techniques*, 1999, pp. 31.1–31.6.

[24]   Miceli, W.J., and L. A. Gross, "Test Results from the AN/APY-6 SAR/GMTI Surveillance, Tracking and Targeting Radar," *Proc. IEEE Radar Conference*, 2001, pp. 13–17.

[25]   Yadin, E., "Evaluation of Noise and Clutter Induced Relocation Errors," *IEEE International Radar Conference*, 1995, pp. 650–655.

[26]   Wang, G., et al., "Detection, Location and Imaging of Fast Moving Targets Using Multi-Frequency Antenna Array SAR," *SPIE Conference on Algorithms for Synthetic Aperture Radar Imagery VII*, Vol. 4053, 2000, pp. 428–439.

[27]   Stockburger, E. F., and D. N. Held, "Interferometric Moving Ground Target Imaging," *IEEE 1995 International Radar Conference*, 1995, pp. 438–443.

[28]   Friedlander, B., and B. Porat, "VSAR: a High Resolution Radar System for Detection of Moving Targets," *IEE Proc. Radar, Sonar and Navigation*, Vol. 144, No. 4, 1997, pp. 205–218.

[29]   Barbarossa, S., and A. Farina, "Detection and Imaging of Moving Objects with Synthetic Aperture Radar—Part 2: Joint Time-Frequency Analysis by Wigner-Ville Distribution," *IEE Proc.-F*, Vol. 139, No. 1, 1992, pp. 89–97.

[30]   Sklansky, J., "On the Hough Technique for Curve Detection," *IEEE Trans. on Computers*, Vol. 27, No. 10, 1978, pp. 923–926.

[31]   Barbarossa, S., and A. Zanalda, "A Combined Wigner-Ville and Hough Transform for Cross-Terms Suppression and Optimal Detection and Parameter Estimation," *IEEE International Conference on Acoustics, Speech and Signal Processing (ICASSP)*, Vol. 5, 1992, pp. 173–176.

[32]   Barbarossa, S., and A. Farina, "Space-Time-Frequency Processing of Synthetic Aperture Radar Signals," *IEEE Trans. on Aerospace and Electronic Systems*, Vol. 30, No. 2, 1994, pp. 341–358.

[33]   Hough, P. V. C., "Method and Means for Recognizing Complex Patterns," *U.S. Patent 3,069,654*, December 1962.

[34]   Duda, R. O., and P. E. Hart, "Use of the Hough Transform to Detect Lines and Curves in Pictures," *Commun. Ass. Comput. Mach.*, Vol. 15, No. 1, 1972, pp. 11–15.

[35]   Chen, V. C., and S. Qian, "Joint Time-Frequency Transform for Radar Range-Doppler Imaging," *IEEE Trans. on Aerospace and Electronic Systems*, Vol. 34, No. 2, 1998, pp. 486–499.

[36]   Chen, V. C., and H. Ling, "Joint Time-Frequency Analysis for Radar Signal and Image Processing," *IEEE Signal Processing Magazine*, Vol. 16, No. 2, 1999, pp. 81–93.

8

Time-Frequency Analysis
of Micro-Doppler Phenomenon

When a radar transmits a signal to a target, the transmitted signal interacts with the target and returns back to the radar. The change in the property of the returned signal reflects characteristics of interest in the target. When the transmitted signal of a coherent Doppler radar hits a moving target, the wavelength of the signal will be changed, and the carrier frequency of the signal will be shifted, known as the Doppler effect. The Doppler frequency shift reflects the velocity of the moving target. When the target is moving away from the radar, the return signal will have a longer wavelength or negative Doppler shift; if the target is moving towards the radar, the return signal will have a shorter wavelength or positive Doppler shift. As we mentioned in Chapter 1, coherent radar systems can preserve phase information of the returned signal. From the obtained phase function of time, by taking the time derivative of the phase, Doppler shifts induced by targets' motions can be found.

An important application of the coherent radar is target identification. A target can be identified based on its signature (i.e., a distinctive characteristic that indicates identity of a target). Signature generated from radar targets' returns is called the radar signature of the target. Radar signature can be in the time domain, such as a range profile of a target, or in the frequency domain, such as the spectrum of a range profile. In this chapter, we will introduce radar signature in a joint time-frequency domain. The time-frequency signature is especially useful to catch time-dependent frequency characteristics.

Mechanical vibration or rotation of structures on a target may induce frequency modulation on the returned radar signal and generates sidebands about the Doppler frequency shift of the target's body. The modulation due to vibrations is called the micro-Doppler phenomenon, originally introduced in coherent laser radar or ladar systems [1, 2]. Because ladar uses the same principle as radar, we adopt it here for radar applications. The modulation induced by rotations, which can be seen as a special case of vibration, can also be interpreted as micro-Doppler. The micro-Doppler phenomenon can be regarded as a characteristic of the interaction between the vibrating or rotating structures and the target body. The change of the properties of back-scattering enables us to determine some properties of the target, and thus the target of interest can be identified based on its micro-Doppler signature. The micro-Doppler provides an additional piece of information for target recognition that is complementary to existing recognition methods [3].

In this chapter, we examine the use of time-frequency analysis for the micro-Doppler phenomenon. In Section 8.1, we analyze micro-Doppler induced by a vibrating scattering center, illustrate time-frequency signature of a vibrating point-scatterer using radar-measured data, and demonstrate time-frequency signature of micro-Doppler generated by the swinging arms of a walking man. In Section 8.2, we introduce micro-Doppler induced by rotation structures and examine the important example of rotor blades. We review the time-domain and the frequency-domain signatures of radar returns from rotor blades, and introduce the time-frequency signatures of rotor blades using both simulated and measured radar data.

8.1 Vibration-Induced Micro-Doppler

In a coherent radar, the phase of a signal returned from a target is sensitive to the variation in range. A half-wavelength change in range can cause 360-degree phase change. It is conceivable that the vibration of a reflecting surface may be measured with the phase change. Thus, the Doppler frequency shift, which represents the change of phase with time, can be used to detect vibrations of structures on a target [3]. These characteristics of vibration are useful for detection and recognition of targets.

Figure 8.1 illustrates a radar located at the origin of the radar coordinate system (X, Y, Z) and a point-scatterer P vibrating about a center point Q that is also the origin of reference coordinates (x, y, z) translated from (X, Y, Z) and at a distance R_0 from the radar. We assume that the center

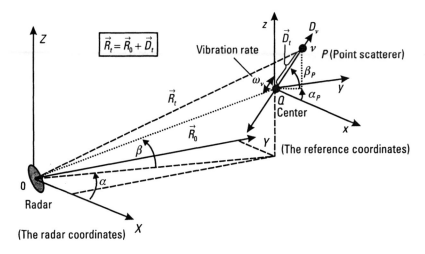

Figure 8.1 Geometry of the radar and a vibrating reflector.

point Q is stationary with respect to the radar. If the azimuth and elevation angles of the point Q with respect to the radar are α and β, respectively, the point Q is located at $(R_0 \cos\beta \cos\alpha, R_0 \cos\beta \sin\alpha, R_0 \sin\beta)$ in the (X, Y, Z) coordinates. We assume that the scatterer P is at a distance D_t from the center point Q and that the azimuth and elevation angles of the scatterer P with respect to the center point Q are α_P and β_P, respectively. Then, the scatterer P will be located at $(D_t \cos\beta_P \cos\alpha_P, D_t \cos\beta_P \sin\alpha_P, D_t \sin\beta_P)$ in the reference coordinates (x, y, z). Therefore, the vector from the radar to the scatterer P becomes $\vec{R}_t = \vec{R}_0 + \vec{D}_t$ as shown in Figure 8.1. The range from the radar to the scatterer P can be expressed as

$$R_t = |\vec{R}_t| = [(R_0 \cos\beta \cos\alpha + D_t \cos\beta_P \cos\alpha_P)^2$$
$$+ (R_0 \cos\beta \sin\alpha + D_t \cos\beta_P \sin\alpha_P)^2 \qquad (8.1)$$
$$+ (R_0 \sin\beta + D_t \sin\beta_P)^2]^{1/2}$$

When $R_0 \gg D_t$, the range is approximately

$$R_t = \{R_0^2 + D_t^2 + 2R_0 D_t [\cos\beta \cos\beta_P \cos(\alpha - \alpha_P) + \sin\beta \sin\beta_P]\}^{1/2}$$
$$\approx R_0 + D_t [\cos\beta \cos\beta_P \cos(\alpha - \alpha_P) + \sin\beta \sin\beta_P] \qquad (8.2)$$

In the case that the azimuth angle α of the center point Q and the elevation angle β_P of the scatterer P are all zero, if $R_0 \gg D_t$ we have

$$R_t = (R_0^2 + D_t^2 + 2R_0 D_t \cos\beta \cos\alpha p)^{1/2} \cong R_0 + D_t \cos\beta \cos\alpha p$$

If the vibration rate of the scatterer in angular frequency is ω_v and the amplitude of the vibration is D_v, then $D_t = D_v \sin\omega_v t$ and the range of the scatterer becomes

$$R(t) = R_t = R_0 + D_v \sin\omega_v t \cos\beta \cos\alpha p \tag{8.3}$$

Thus, the radar received signal becomes

$$s_R(t) = \rho \exp\left\{ j\left[2\pi f_0 t + 4\pi \frac{R(t)}{\lambda} \right] \right\} = \rho \exp\{j[2\pi f_0 t + \Phi(t)]\} \tag{8.4}$$

where ρ is the reflectivity of the point-scatterer, f_0 is the carrier frequency of the transmitted signal, λ is the wavelength, and $\Phi(t) = 4\pi R(t)/\lambda$ is the phase function.

Substituting (8.3) into (8.4) and denoting $B = (4\pi/\lambda)D_v \cos\beta \cos\alpha p$, the received signal can be rewritten as

$$s_R(t) = \rho \exp\left\{ j\frac{4\pi}{\lambda} R_0 \right\} \exp\{j2\pi f_0 t + B\sin\omega_v t\} \tag{8.5}$$

which can be further expressed by the Bessel function of the first kind of order k:

$$J_k(B) = \frac{1}{2\pi} \int_{-\pi}^{\pi} \exp\{j(B\sin u - ku)du\} \tag{8.6}$$

and, thus,

$$s_R(t) = \rho \exp\left(j\frac{4\pi}{\lambda} R_0 \right) \sum_{k=-\infty}^{\infty} J_k(B) \exp[j(2\pi f_0 + k\omega_v)t]$$

$$= \rho \exp\left(j\frac{4\pi}{\lambda} R_0 \right) \{J_0(B) \exp(j2\pi f_0 t) \tag{8.7}$$

$$+ J_1(B) \exp[j(2\pi f_0 + \omega_v)t] - J_1(B) \exp[j(2\pi f_0 - \omega_v)t]$$

$$+ J_2(B) \exp[j(2\pi f_0 + 2\omega_v)t] + J_2(B) \exp[j(2\pi f_0 - 2\omega_v)t]$$

$$+ J_3(B) \exp[j(2\pi f_0 + 3\omega_v)t] - J_3(B) \exp[j(2\pi f_0 - 3\omega_v)t]$$

$$+ \dots \}$$

Therefore, the micro-Doppler frequency spectrum consists of pairs of spectral lines around the center frequency f_0 and with spacing $\omega_v/(2\pi)$ between adjacent lines.

By taking the time-derivative of the phase function in (8.4), the micro-Doppler frequency induced by the vibration is a sinusoidal function of time at the vibration frequency ω_v:

$$f_D = \frac{1}{2\pi}\frac{d\Phi}{dt} = \frac{2}{\lambda}\frac{dR(t)}{dt} \qquad (8.8)$$

$$= \frac{2}{\lambda}D_v\omega_v[\cos\beta\cos\beta_P\cos(\alpha - \alpha_P) + \sin\beta\sin\beta_P]\cos\omega_v t$$

If the azimuth angle and the elevation angle β_P are all zero, we have

$$f_D = \frac{2}{\lambda}D_v\omega_v\cos\beta\cos\alpha_P\cos\omega_v t \qquad (8.9)$$

When the orientation of the vibrating scatterer is along the projection of the radar LOS direction, or $\alpha_P = 0$, and the elevation angle β of the scatterer is also 0, the Doppler frequency change reaches the maximum value of $(2/\lambda)D_v\omega_v$.

Usually, when the vibrating modulation is small, it is difficult to detect the vibration in the frequency domain. Thus, a method that removes the energy of the target's Doppler velocity and keeps only the residual Doppler (i.e., the micro-Doppler), may help distinguish vibration spectrum from other contributions.

8.1.1 Time-Frequency Signature of a Vibrating Scatterer

According to (8.9), when a radar is operating at X-band with a wavelength of 3 cm, a vibration rate at 10 Hz with a displacement of 0.1 cm will induce a maximal micro-Doppler frequency shift of 0.66 Hz as shown in Figure 8.2. The micro-Doppler shift may be detectable with a high frequency-resolution radar. This is illustrated by analyzing a time series collected using an X-band SF radar illuminating two point-scatterers shown in Figure 8.3(a). The two scatterers are separated by a distance of 13.5 m. One scatterer is stationary and the other is vibrating at 1.5 Hz with a displacement of 3 cm [4]. Although the vibration rate is relatively low, according to (8.9) the displacement of 3 cm can still generate larger micro-Doppler frequency shift up to 3 Hz. Figure 8.3(b) is the magnitude of the I and Q returned signals

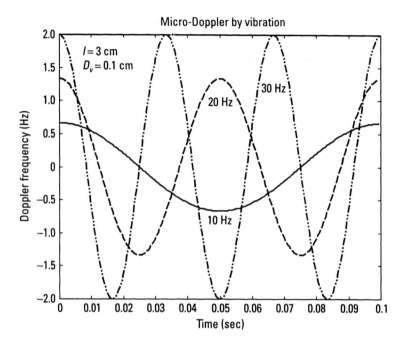

Figure 8.2 Micro-Doppler generated by a vibrating reflector with a vibration rate of 10 Hz, 20 Hz, and 30 Hz, respectively.

from the two scatterers during a short time interval. By taking the Fourier transform of the I and Q signals, the magnitude of its Fourier frequency spectrum has one sharp peak around 0.29 and one peak spread around 0.35 to 0.39 as shown in Figure 8.3(c), where the maximum frequency scale is normalized to 0.5. As we described in Chapter 1, for an SF radar waveform, which sequentially transmits short rectangular pulses with a carrier frequency stepped from one pulse to the next, the pulse compression is performed by the Fourier transform. By taking the Fourier transform, the radar received SF signal can be compressed and appears as a radar "range profile." In other words, the Fourier spectrum in Figure 8.3(c) is actually a "range profile," where the two peaks represent the two point-scatterers located in range. The peak spread in the spectrum indicates that there may be a modulation due to vibrating. However, the width of the peak measured from the spectrum does not reflect the true displacement of the vibrating scatterer because the "range profile" is actually the Fourier spectrum in the frequency-domain.

Figure 8.3(d) shows the time-frequency signature of the radar signal returned from the two scatterers, where the vibration curve can be observed very well. It is obtained by applying a Gabor transform described in Chapter 2.

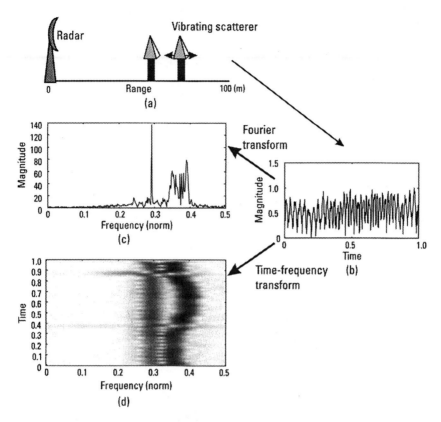

Figure 8.3 An experimental radar-returned signal from two point-scatterers: (a) geometry of the radar and the two scatterers; (b) the magnitude of the radar-returned signal; (c) the Fourier transform of the radar-returned signal; and (d) the time-frequency transform of the returned signal. (Data provided by S. Wong, DREO of Canada.)

From the time-frequency signature we can see that the micro-Doppler of the vibrating scatterer is a time-varying frequency spectrum. From the additional time information, it is possible to estimate the vibration rate of the vibrating scatterer by measuring the period of the vibration curve. However, the maximum displacement of the vibration curve does not reflect the true displacement in range, as mentioned earlier.

8.1.2 An Example of Micro-Doppler Signatures of Moving Targets

Here we demonstrate an example of micro-Doppler signatures of moving targets. The moving target in this example is a walking man with swinging arms. The radar is mounted on the rooftop of a building, and a man is walking towards the building at a speed of about 1.8 m/sec as illustrated in Figure 8.4(a). The radar data has 64 range cells and 1,000 pulses and was collected at a PRF of 800 Hz. Figure 8.4(b) shows radar-range and cross-range images of the walking man generated by 64 range-cells and 128 pulses, where the hot spot in the image indicates the body of the walking man. We also notice that there are smeared lines running across the cross-range direction around the body of the walking man at the range cell 12. If we apply a Gabor transform to the time history data at the range cell 12, the body Doppler shift and the micro-Doppler signature of the swinging arms can be clearly detected in the time-frequency domain [4]. As shown in Figure 8.4(c), the Doppler shift of one arm is higher and the other is lower than the body Doppler frequency shift. Figure 8.4(d, e) shows the result with

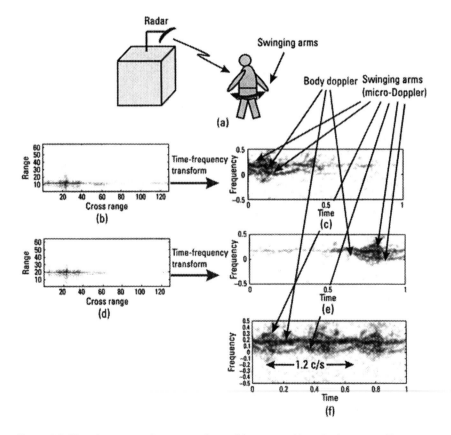

Figure 8.4 Time-frequency signatures of a walking man with swinging arms. (Data provided by H. Holt, Norden Systems, Northrop Grumman.)

the same data but in a different short-time interval and at the range cell 20. Superposition of the time-frequency signatures over several range cells, which correspond to different walking steps, gives a full time-frequency signature of the walking man as shown in Figure 8.4(f). We can see that the body's Doppler shift is almost constant but the arm's micro-Doppler shift becomes time-varying and has a sinusoidal-like curve. Again, from the additional time information, the swinging rate of the arm can be estimated and is about 1.2 cycle/sec in this example.

8.2 Rotation-Induced Micro-Doppler

Modulation induced by rotation structure (such as rotor blades of a helicopter, propellers of an aircraft, or rotating antennas on a ship or an aircraft) can be regarded as a unique signature of the target. The micro-Doppler signature becomes an important feature for identifying the target of interest. Here, we examine micro-Doppler induced by rotor blades of a helicopter, show the time-domain and the frequency-domain signatures of the micro-Doppler, and introduce the time-frequency signature of the micro-Doppler induced by rotor blades.

8.2.1 Rotor Blade Motion

In a helicopter, the main rotor blades, the tail rotor blades, and the hub have unique signatures suitable for target identification. Generally, radar returns from a helicopter are back-scattered from the fuselage, the rotor blades, the tail blades, the hub, and other structures on the helicopter. The motion of the rotor blades depends on the interdependent coupling between the aerodynamics and the rotor dynamics [5]. Each blade is a rotating aerofoil having bending, flexing, and twisting. The radar cross section of a segment in the blade depends upon its distance from the center of rotation, its angular position, and the aspect angle of the rotor [6, 7].

Because we are especially interested in electromagnetic back-scattering from the main rotating blades of a helicopter, for simplicity, the rotor blade is modeled as a rigid, homogeneous, and linear rod rotating about a fixed axis with a constant rotation rate. No flapping, lagging, and feathering are considered for the calculation of electromagnetic back-scattering.

8.2.2 Radar Returns from Rotor Blades

As illustrated in Figure 8.5, the radar is located at the origin of the radar coordinates (X, Y, Z); reference coordinates (x, y, z) are translated from

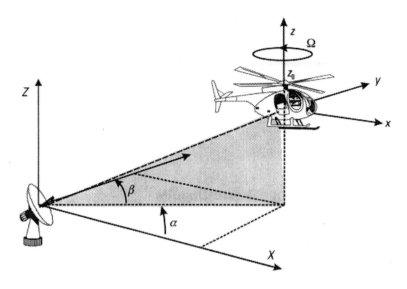

Figure 8.5 Geometry of the radar and the helicopter.

the radar coordinates and their origin is at the geometric center of the helicopter. The rotor blades are on the plane $(x, y, z = z_0)$ and rotate about the z-axis with a rotation rate of Ω. The azimuth and the elevation angles of the helicopter with respect to the radar are α and β, respectively.

Let us begin with a simple case where $\alpha = \beta = 0$ and $z_0 = 0$ as shown in Figure 8.6. Assume a scatterer P at (x_0, y_0) on a rotor blade rotates about a center point Q with a rotation rate of Ω. The distance from the scatterer to the center point is l, and the distance between the radar and the center point is R_0, which may be a function of time if the target is moving. Assume that at $t = 0$ the initial rotation angle of the scatterer in the blade is θ_0, and at t the rotation angle becomes $\theta_t = \theta_0 + \Omega t$ and the scatterer is rotated to

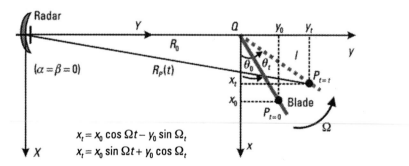

Figure 8.6 Geometry of the radar and a rotor blade when $\alpha = \beta = 0$.

(x_t, y_t). Because we assume both the radar and the rotor are on the same plane, at time t the range from the radar to the scatterer can be derived as

$$R_P(t) = [R_0^2 + l^2 + 2lR_0 \sin(\theta_0 + \Omega t)]^{1/2} \qquad (8.10)$$
$$\cong R_0 + V_R t + l \sin \theta_0 \cos \Omega t + l \cos \theta_0 \sin \Omega t$$

where V_R is the radial velocity of the helicopter and $(l/R_0)^2 \to 0$ for far field. Thus, the radar received signal becomes

$$s_R(t) = \exp\left\{ j \left[2\pi f_0 t + \frac{4\pi}{\lambda} R_P(t) \right] \right\} = \exp\{ j[2\pi f_0 t + \Phi_P(t)] \}$$
$$(8.11)$$

where $\Phi_P(t) = 4\pi R_P(t)/\lambda$ is the phase function.

If the elevation angle β and the height of rotor blades z_0 are not zero, then the phase function should be modified as

$$\Phi_P(t) = \frac{4\pi}{\lambda} [R_0 + V_R t + \cos \beta (l \sin \theta_0 \cos \Omega t + l \cos \theta_0 \sin \Omega t) + z_0 \sin \beta]$$
$$(8.12)$$

The returned signal from the scatterer P in the rotor blade becomes

$$s_R(t) = \exp\left\{ j \frac{4\pi}{\lambda} [R_0 + V_R t + z_0 \sin \beta] \right\} \qquad (8.13)$$
$$\cdot \exp\left\{ j 2\pi f_0 t + \frac{4\pi}{\lambda} l \cos \beta \sin(\Omega t + \theta_0) \right\}$$

Let $\theta_0 = 0$, denote B $= (4\pi/\lambda)l \cos \beta$, and reduce (8.13) to (8.5). Again, it can be expressed by the Bessel function of the first kind. Similar to (8.7), the micro-Doppler of a scatterer on the blade consists of pairs of spectral lines around the center frequency f_0 and with spacing $\Omega/(2\pi)$ between adjacent lines.

After compensating the motion and removing the constant phase term in (8.13), the baseband signal returned from the scatterer P becomes

$$s_B(t) = \exp\left\{ j \frac{4\pi}{\lambda} l \cos \beta \sin(\Omega t + \theta_0) \right\} \qquad (8.14)$$

By integrating (8.14) over the length of the blade L, the total baseband signal becomes the following [6]:

$$s_L(t) = \int_0^L \exp\left\{ j\frac{4\pi}{\lambda} l \cos\beta \sin(\Omega t + \theta_0) \right\} dl \qquad (8.15)$$

$$= L \exp\left\{ j\frac{4\pi}{\lambda}\frac{L}{2} \cos\beta \sin(\Omega t + \theta_0) \right\} \text{sinc}\left\{ \frac{4\pi}{\lambda}\frac{L}{2} \cos\beta \sin(\Omega t + \theta_0) \right\}$$

For a rotor with N blades, there will be N different initial rotation angles

$$\theta_k = \theta_0 + k2\pi/N, \ (k = 0, 1, 2, \ldots N - 1) \qquad (8.16)$$

and the total received signal becomes

$$s_\Sigma(t) = \sum_{k=0}^{N-1} s_{Lk}(t) \qquad (8.17)$$

$$= \sum_{k=0}^{N-1} L \, \text{sinc}\left\{ \frac{4\pi}{\lambda}\frac{L}{2} \cos\beta \sin(\Omega t + \theta_0 + k2\pi/N) \right\} \exp\{j\Phi_k(t)\}$$

where

$$\Phi_k(t) = \frac{4\pi}{\lambda}\frac{L}{2} \cos\beta \sin(\Omega t + \theta_0 + k2\pi/N) \quad (k = 0, 1, 2, \ldots N - 1)$$

$$(8.18)$$

8.2.3 Time-Domain Signatures of Rotation-Induced Modulations

Rotor blades in a helicopter are in rotational motion that will impart a periodic modulation on radar returned signals as shown in (8.17). The rotation-induced Doppler shifts relative to the Doppler shift of the fuselage (or body) occupy unique locations in the frequency domain. The modulation in the frequency domain as well as the time-domain signal have been used as radar signatures for target identification [8, 9].

The time-domain signature of rotor blades is defined by the magnitude in (8.17)

$$|s_\Sigma(t)| = \left| \sum_{k=0}^{N-1} L \operatorname{sinc}\left\{ \frac{4\pi}{\lambda} \frac{L}{2} \cos\beta \sin(\Omega t + \theta_0 + k2\pi/N) \right\} \exp\{j\Phi_k(t)\} \right|$$

(8.19)

where $\Phi_k(t)$ is defined by (8.18).

Assume a radar is operating at S-band with a wavelength of 0.1m, and a helicopter has 2 rotor blades at a constant rotation rate of 5 revolutions/sec (rev/sec). If the distance of the blade roots from the rotor center is 0.3m, the distance of the blade tips from the rotor center is 6.7m, and the elevation angle of the rotor $\beta = 0$, the time-domain signature of the rotor blades in (8.19) is shown in Figure 8.7(a). The frequency spectrum of the same signal is shown in 8.7(b), which we will discuss in Section 8.2.4. The rotor blade's return has a short flash when the blade has specular reflection at the advancing as well as at the receding point of rotation [9]. The interval between flashes is related to the rotation rate of the rotor. The duration of the flash is determined by the length of the blade L, the wavelength λ, and the rotation rate Ω as described by the sinc function in (8.19).

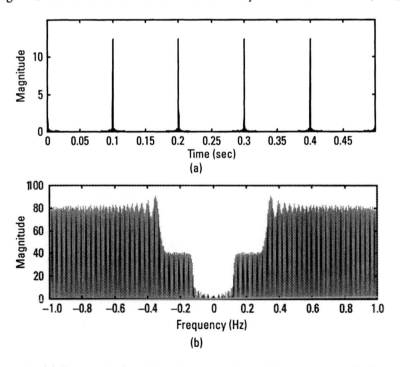

Figure 8.7 (a) The magnitude of the radar-returned signal from two rotor blades; and (b) the frequency spectrum of the radar-returned signal.

For longer blade length and at a shorter wavelength, the duration of the flash is shorter. Because the number of blades is 2 and the rotation rate is 5 rev/sec, there are 5 flashes in 0.5 sec, and the interval between flashes is 0.1 sec as shown in Figure 8.7(a).

Figure 8.8(a) shows a simple Computer-Aided Design (CAD) model of a helicopter with four rotor blades. An electromagnetic prediction code *Xpatch* was used to simulate radar returns from the helicopter model [10, 11]. The body of the helicopter is made by two conducting-boxes with different sizes. The length of the rotor blade is 3m. In the simulation, the radar scattering is observed at 128 frequency steps from 1.75 GHz to 2.25 GHz and 128 target aspect looks from 158 to 170 degrees. The elevation angle of the helicopter is 20 degrees. The blade position is articulated between 0 and 360 degrees during the 128 angular looks. The rotational motion of the helicopter's body is 12 rev/m or 0.2 rev/sec. The rotation rate of the rotor blade is 360 rev/m or 6 rev/sec [12]. Figure 8.8(b) is the standard radar image of the helicopter in the range versus cross-range domain by taking the 2D Fourier transform of the *Xpatch*-generated data. We can see that the body of the helicopter has a small extent around the center of the cross-range while the fast rotating rotor blades exhibit strong smeared lines running across the cross-range and overlapping with the body of the helicopter. The time-domain signature of the simulated helicopter is shown in Figure 8.9(a), where the radar signal includes returns from both the rotor blades and the helicopter body. The frequency spectrum of the same simulated helicopter is shown in 8.9(b), which will be discussed in Section 8.2.4.

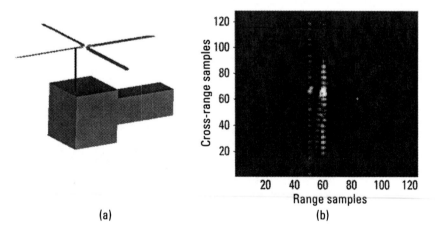

(a)   (b)

Figure 8.8 (a) A simplified CAD model of a helicopter with four rotor blades; and (b) the radar range and cross-range image of the helicopter.

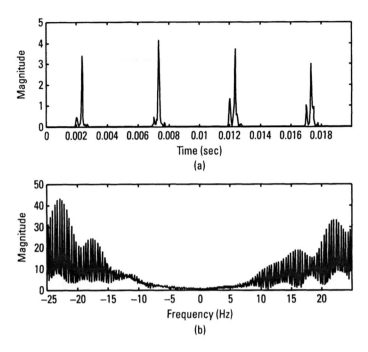

Figure 8.9 (a) The magnitude of the radar-returned signal from the simplified CAD model of a helicopter; and (b) the frequency spectrum of the radar-returned signal.

Figure 8.10(a) shows a similar time-domain signature of measured helicopter data. The overlapped range profiles illustrated in Figure 8.11 show separated flash peaks of rotor blades. It also shows overlapped peaks of returns from the body of the helicopter. Unlike the flash peaks, these overlapped body peaks are located at a fixed range cell. The corresponding frequency spectrum of the measured helicopter data is shown in 8.10 (b), which will be discussed in Section 8.2.4.

As was defined in Chapter 1, for each transmitted pulse, the time-domain signature is actually a range profile. From only one range profile, it is difficult to observe the rotation feature of rotor blades. However, owing to the frequency modulation induced by rotational motions, the rotation feature may be observed from the frequency-domain signature in its Doppler spectrum. To better observe the rotation feature of rotor blades, we use a sequence of range profiles by transmitting a sequence of pulses and rearranging them into a 2D (the range versus the dwell time) array matrix. At each range cell, the time history series across the range profiles provides a better way to observe the rotation behavior during a longer time period. This will be further discussed in Section 8.2.5.

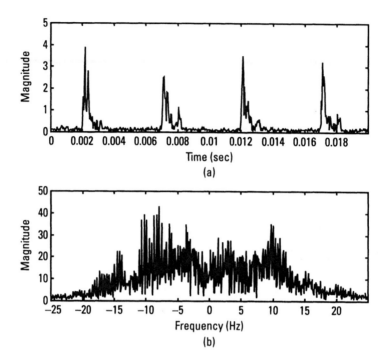

Figure 8.10 (a) The magnitude of the radar-returned signal from a helicopter; and (b) the frequency spectrum of the radar-returned signal.

8.2.4 Frequency-Domain Signatures

Because the time-derivative of the phase is the frequency, by taking the time-derivative of the phase function $\dfrac{4\pi}{\lambda}\dfrac{L}{2}\cos\beta\sin(\Omega t + \theta_0)$ in (8.15), the Doppler frequency shift induced by a rotor blade becomes

$$f_D(t) = \frac{L}{\lambda}\Omega\cos\beta(-\sin\theta_0\sin\Omega t + \cos\theta_0\cos\Omega t) \qquad (8.20)$$

If the initial rotation angle $\theta_0 = 0$, then we have $f_D(t) = \dfrac{L}{\lambda}\Omega\cos\beta\cos\Omega t$. We can clearly see that the Doppler frequency is modulated by the rotation rate Ω through $\cos\Omega t$. The frequency spectrum of the received signal can be directly obtained by taking the Fourier transform of (8.15).

For a rotor with N blades, the frequency spectrum of the total received baseband signal in (8.17) can be expressed as follows [6]:

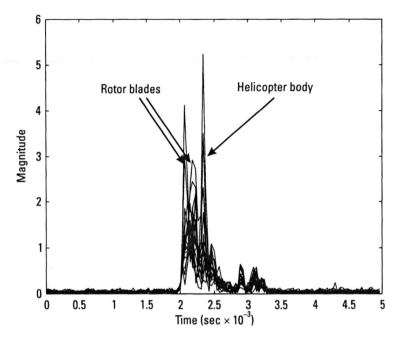

Figure 8.11 Overlapped individual range profiles of the measured radar data of a helicopter.

$$S_{\Sigma}(f) = \sum_{m=-M}^{M} C_m \delta(f - mN\Omega) \qquad (8.21)$$

where C_m is a scale factor determined by Bessel functions of the first kind with parameters of λ, L, β, N, and m, and M is the index number of the most significant sideband. Because we have compensated the translational motion of the rotor blades and removed the carrier frequency and the constant phase term, the residual frequency spectrum reflects only the micro-Doppler shifts induced by rotor blades relative to the zero frequency.

Figure 8.7(b) shows the frequency spectrum of radar returns from only the rotor blades and based on the baseband time-domain signal derived in (8.17). The Doppler modulation about center (zero) frequency can be seen. The lower cutoff frequency is determined by the distance between the rotor center and the blade roots. For the *Xpatch*-simulated radar data described in Section 8.2.3, the radar return is not only from rotor blades, but also from the body of the helicopter. The frequency spectrum of the data is shown in Figure 8.9(b), where the Doppler modulation about the center (zero) frequency can also be seen. For the measured radar data described in

Section 8.2.3, radar returns also include returns from structures other than rotor blades of the helicopter. The frequency signature is shown in Figure 8.10(b), where the Doppler frequency modulation about the center (zero) frequency can also be seen. However, due to noise spectrum, the magnitude of the spectrum around the center frequency is higher.

Frequency-domain signatures provide information about frequency modulations by either rotating blades or other rotating or vibrating structures. Because of the lack of time information, it is not easy to tell the rotation rate from the frequency spectrum alone. Therefore, the time-frequency signature that provides time-dependent frequency information is more useful as an additional information for target identification complementary to existing time-domain or frequency-domain methods.

8.2.5 Time-Frequency Signatures

Figure 8.12 illustrates a stack of radar range profiles of a helicopter. In each range profile, we assume that there are returns from four rotor blades and one return from the body of the helicopter. Because of the blades' rotation, the amplitudes of the blades' returns change from one profile to the next as seen in Figure 8.12. Thus, four blades have flashes at different times.

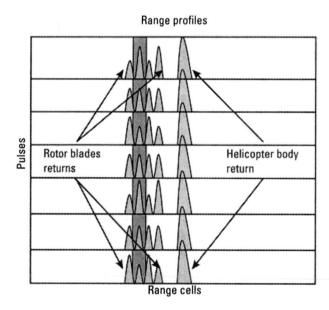

Figure 8.12 A stack of radar range profiles returned from a helicopter.

However, the amplitude of the return from the body is relatively stable. When we apply the Fourier transform to form an image of the helicopter, the image of the body is relatively clear but the image of the rotor blades is smeared along the cross-range domain.

Although range profiles can provide features, such as the location and the reflectivity of a scatterer, a better way to extract useful features of rotor blades may not be from the range profile. Instead, the data across range profiles at a range cell where the smearing in the cross-range occurs may be used. To extract features of rotor blades, we observe the time-frequency behavior of rotor blades by taking a time-frequency transform of the data across range profiles.

Figure 8.13 shows the time-frequency signature of the returned signal from the *Xpatch*-simulated helicopter shown in Figure 8.8, where the characteristics of the rotating blades can be seen more clearly in the joint time-frequency domain. The strong time-frequency distribution along the horizontal line about the center frequency is due to returns from the helicopter's body. The strong time-frequency distribution along slope-lines about

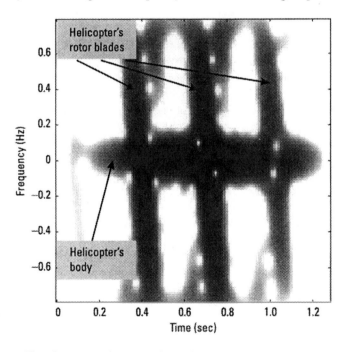

Figure 8.13 Time-frequency signature of the simulated radar-returned signal from the simplified CAD model of a helicopter.

the vertical direction is due to returns from the rotating blades. Because time information is now available, the rotation rate of the blades can be measured from their time-frequency signatures [13].

References

[1]   Parker, K. J., R. M. Lerner, and S. R. Huang, "Method and Apparatus for Using Doppler Modulation Parameters for Estimation of Vibration Amplitude," *U.S. Patent 5,086,775*, February 11, 1992.

[2]   Zediker, M. S., R. R. Rice, and J. H. Hollister, "Method for Extending Range and Sensitivity of a Fiber Optic Micro-Doppler Ladar System and Apparatus Therefor," *U.S. Patent 6,847,817*, December 8, 1998.

[3]   Sommer, H., and J. Salerno, "Radar Target Identification System," *U.S. Patent 3,614,779*, October 19, 1971.

[4]   Chen, V. C., "Analysis of Radar Micro-Doppler Signature with Time-Frequency Transform," *Proc. Tenth IEEE Workshop on Statistical Signal and Array Processing*, August 2000, pp. 463–466.

[5]   Leishman, J. G., *Principles of Helicopter Aerodynamics*, Cambridge University Press, 2000.

[6]   Martin, J., and B. Mulgrew, "Analysis of the Theoretical Radar Return Signal from Aircraft Propeller Blades," *IEEE International Radar Conference*, 1990, pp. 569–572.

[7]   Fliss, G. G., and D. L. Mensa, "Instrumentation for RCS Measurements of Modulation Spectra of Aircraft Blades," *IEEE National Radar Conference*, 1986, pp. 95–99.

[8]   Bullard, B. D., and P. C. Dowdy, "Pulse Doppler Signature of a Rotary-Wing Aircraft," *IEEE AES Systems Magazine*, May 1991, pp. 28–30.

[9]   Misiurewicz, J., K. Kulpa, and Z. Czekala, "Analysis of Recorded Helicopter Echo," *Proc. IEE Radar*, 1997, pp. 449–453.

[10] Andersh D. J., et al., "Xpatch: A High Frequency Electromagnetic-Scattering Prediction Code and Environment for Complex Three-Dimensional Objects" *IEEE Antennas and Propagation Magazine*, Vol. 6, February 1994, pp. 65–69.

[11] Bhalla, R., and H. Ling, "A Fast Algorithm for Simulating Doppler Spectra of Targets with Rotating Parts Using the Shooting and Bouncing Ray Technique," *IEEE Trans. on Antennas and Propagation*, Vol. 46, September 1998, pp. 1389–1391.

[12] Wang, Y., H. Ling, and V. C. Chen, "Application of Adaptive Joint Time-Frequency Processing to ISAR Image Enhancement and Doppler Feature Extraction for Targets with Rotating Parts," *SPIE Proc. on Radar Processing, Technology, and Application III*, Vol. 3462, pp. 156–163, 1998.

[13] Chen, V.C., "Radar Signatures of Rotor Blades," *SPIE Proc. on Wavelet Applications VIII*, Vol. 4391, 2001.

9

Trends in Time-Frequency Transforms for Radar Applications

The Fourier transform has been widely used in radar signal and image processing. When radar signals exhibit time-varying behavior, a transform that represents the intensity or energy distribution in the joint time and frequency domain is most desirable. In Chapters 3 through 8, we have shown that the time-frequency transform is indeed a useful tool for the detection of weak signals buried in noise, for analyzing radar back scattering, for focusing the image of maneuvering targets, for motion compensation, and for micro-Doppler analysis. We discuss here some current trends in time-frequency transforms for radar applications.

9.1 Applications of Adaptive Time-Frequency Transforms

Parametric time-frequency transforms, also called model-based or adaptive transforms, make use of time-frequency basis functions and adaptively select the parameters of the basis functions to decompose a signal into the joint time-frequency domain [1–5].

In Sections 2.1.3 and 6.2.1, we have discussed some radar applications of adaptive time-frequency transforms. Gabor function with three parameters (time center, frequency center, and time-frequency extent) and linear chirp function were used as basis functions. Their parameters were then selected adaptively by searching for the maximum projection of the signal onto all possible bases in a dictionary of basis functions.

When the basis set is well matched to the analyzed signal, the performance of adaptive time-frequency algorithms can be very good. Compared to the nonparametric methods, however, adaptive algorithms are usually computationally quite expensive, especially if the number of bases in the dictionary is large. A fast scheme is proposed in [5] for simple chirplet functions to overcome this problem. Fast algorithms for more general basis functions are still very much needed.

9.2 Back-Scattering Feature Extraction

Chapter 4 showed that time-frequency analysis is a very useful tool in unveiling the underlying scattering phenomenology in radar back-scattering data. This led to a better understanding of the various scattering mechanisms in complex targets. Without time-frequency analysis, these mechanisms were not easy to interpret from measured or computed data, nor could they be easily studied from analytical solutions of Maxwell's equations due to target complexity. This line of research should be continued to help build up a more comprehensive knowledge base for complex shapes and exotic materials. For target identification applications, work to incorporate the extracted time-frequency features to improve the performance of existing classifiers is also worthwhile pursuing [6, 7]. Finally, a more thorough understanding of the scattering phenomenology from the electromagnetics point of view will allow us to devise better basis functions in model-based time-frequency algorithms. This can lead to physics-based signal processing algorithms that significantly out-perform existing time-frequency tools.

Another potential application area where time-frequency analysis might play a useful role is the suppression of clutter and propagation effects. For instance, ocean surface often gives rise to large clutter that makes the detection and classification of small floating targets a very difficult task [8]. In ground penetrating radar for detecting buried objects, the dispersive effects due to wave propagation through soil can lead to significant image distortion [9]. In foliage penetration (FOPEN) applications, the two-way propagation through tree canopies will be an important factor on how well hidden targets can be detected [10]. It would be fruitful to exploit the difference in target, clutter, and propagation channel characteristics in the context of time-frequency space to achieve clutter suppression, propagation effect removal, and target feature enhancement. Some work along this line on SAR clutter suppression using wavelets has been reported in [11].

9.3 Image Formation

As shown in Chapter 5, time-frequency-based image formation has been successfully applied to radar image formation, especially in the case of blurred Fourier images. The time-frequency distribution series and WVD have been investigated for time-frequency-based image formation [12–14]. Continual search for fast time-frequency transforms with high time-frequency resolution and low cross-term interference is still needed. Analysis of the image resolution provided by different time-frequency transforms is another issue to be further studied.

Compared to the Fourier transform for image formation, the time-frequency-based image formation shows its capability for improving the signal-to-noise ratio. How much SNR improvement can be obtained with time-frequency transforms is still an open issue. A quantitative analysis of the SNR improvement for radar image formation should be further studied.

9.4 Motion Compensation

In Chapter 6, the use of time-frequency analysis for achieving ISAR motion compensation has been discussed. An adaptive time-frequency procedure was used to extract the phases of the prominent point-scatterers on the target. The extracted phase information was then used in conjunction with the PPP model to remove the higher-order motion errors in the radar data. In this procedure, the phase of the resulting focused image is preserved and the Doppler resolution offered by the full coherent processing interval can be achieved.

As in all model-based time-frequency approaches, the model must be well matched to the actual motion physics in order to achieve good performance. When more complex motion dynamics are involved, more sophisticated models should be investigated. For example, for air targets undergoing fast maneuvers or ships on rough seas, the rotational motion of the target may not be confined to a 2D plane. Section 6.4 discussed how the existence of such chaotic motions could be detected using time-frequency processing. However, the question of how motion compensation can be achieved under such situations remains a very challenging research topic [15], and much more work is warranted.

Another important issue is the computational speed of the time-frequency-based procedure for motion compensation. One drawback of the adaptive time-frequency method is the computational burden associated with the exhaustive search procedure for the motion parameters. This problem becomes especially severe when higher-order motions are involved. Recently, genetic algorithms [16] have been attempted as a way to reduce the computational complexity and speed up the search procedure.

9.5 Moving Target Detection

In Chapter 7, we discussed SAR imaging of moving targets and how to detect moving targets in clutter. Especially in a high clutter environment, it is difficult to detect moving targets by using conventional methods. In [17], the WVD was used to detect a moving target and estimate motion parameters. Combined with the Hough transform as we described in Section 7.4.1, the WVD works well for single target detection, but not for multiple targets owing to the cross-term interference. A combined Wigner-Hough transform extended to the analysis of multicomponent LFM signals was suggested in [18–20].

Another approach to detecting multiple LFM signals in clutter is the fractional Fourier transform (FRFT) [21]. For any real angle α, the FRFT of a signal $s(t)$ is defined by

$$FRFT_\alpha(u) = \qquad\qquad\qquad\qquad\qquad\qquad\qquad (9.1)$$

$$\begin{cases} \left(\dfrac{1 - j\cot\alpha}{2\pi}\right)^{1/2} e^{j\frac{u^2}{2}\cot\alpha} \displaystyle\int_{-\infty}^{\infty} s(t) e^{j\frac{t^2}{2}\cot\alpha} e^{-jut\csc\alpha} dt & (\alpha \neq n\pi) \\[3mm] s(u) & (\alpha = n2\pi) \\[2mm] s(-u) & (\alpha + \pi = n2\pi) \end{cases}$$

where n is an integer.

The FRFT depends on an angle parameter α and can be interpreted as a rotation of the time-frequency plane by the angle α. In [22] relationships between the FRFT and the WVD and the STFT are derived in a simple and natural form. For the WVD, the relationship is

$$WVD(t, \omega) = 2\exp\{2j\omega't\} \int_{-\infty}^{\infty} FRFT_\alpha(z)FRFT_\alpha^*(2\omega' - z)\exp\{-2jt'z\}dz$$

$$(9.2)$$

where $\omega' = t\cos\alpha + \omega\sin\alpha$ and $t' = t\sin\alpha + \omega\cos\alpha$. This means the WVD of $FRFT_\alpha$ is the WVD of the signal $s(t)$ rotated by an angle $-\alpha$. For the STFT, the relationship is

$$STFT^M(t,\,\omega) = \frac{1}{(2\pi)^{1/2}} \exp\left\{\frac{\omega't'}{2}\right\} \int_{-\infty}^{\infty} FRFT_\alpha(z)\,W_\alpha^*(\omega' - z)\,\exp\{-jt'z\}dz$$

(9.3)

where $W_\alpha(\cdot)$ is the frequency window function with arguments $(\omega',\,t')$ and $STFT^M$ is a modified STFT defined by

$$STFT^M(t,\,\omega) = \frac{1}{(2\pi)^{1/2}} \exp\left\{\frac{\omega t}{2}\right\} \int_{-\infty}^{\infty} s(t')w^*(t - t')\,\exp\{-j\omega t'\}dt'$$

(9.4)

where $w(\cdot)$ is the time window function. Again, the modified STFT of $FRFT_\alpha$ is a rotated version of the modified STFT of the signal $s(t)$.

Therefore, the time-frequency transform of the FRFT of a signal is the rotated time-frequency transform of the signal by an angle α. When $\alpha = \pi/2$ the FRFT is equivalent to the Fourier transform, and when $\alpha = 0$ it is an identity operator (i.e., the FRFT of a signal is the signal itself). Since the FRFT is just a generalized Fourier transform [23], it is a linear transform and has no cross-term interference. The FRFT has been applied to many areas including signal processing [24–26] and analysis of multiple LFM signals [27].

For a given LFM (or linear chirp signal) with an unknown chirp rate as shown in the joint time-frequency domain in Figure 9.1, the projection of the signal onto the Fourier frequency f-domain is its frequency spectrum, and its energy is spread in frequency. However, if we select a proper rotation angle α, the $(t,\,f)$ coordinates rotate to a new set of $(t',\,\omega')$ coordinates called the fractional Fourier domain, where the projection of the LFM signal onto the fractional Fourier u-domain is the fractional Fourier spectrum. The fractional Fourier spectrum of the signal is highly concentrated in fractional frequency (ω'-axis), and the centroid of the LFM signal can also be determined by the position of its energy peak in fractional Fourier frequency. Thus, with the FRFT and by searching rotation angles, LFM signals generated by moving targets can be detected.

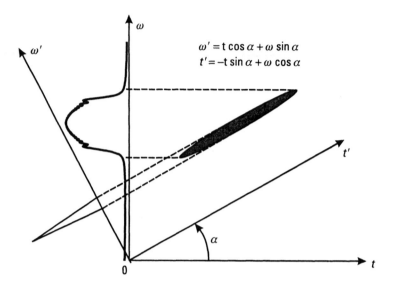

$$\omega' = t \cos \alpha + \omega \sin \alpha$$
$$t' = -t \sin \alpha + \omega \cos \alpha$$

Figure 9.1 Illustration of the Fourier transform and the FRFT of an LFM signal.

A robust FRFT called the generalized-marginal time-frequency distribution, which is a bilinear transform with an angle parameter α in its kernel function, was proposed in [28] and has potential applications for the detection of multiple LFM signals in a strong clutter environment.

9.6 Micro-Doppler Analysis

Micro-Doppler phenomenon has been observed in the frequency domain for those targets that have vibrating or rotating structures. As discussed in Chapter 8, the joint time-frequency representation of the micro-Doppler can provide time information such as the vibration rate or the rotation rate. Unfortunately, so far there are not many publications on the subject of time-frequency analysis of micro-Doppler phenomena.

In Chapter 8, we demonstrated examples of the micro-Doppler phenomenon and its signature in the joint time-frequency domain. Some recent studies on extraction of micro-Doppler features can be found in [29–31].

In [29] an X-band radar is used to capture the natural resonance frequency of large moving vehicles such as a tractor trailer. As described in [29], the life of a highway bridge can be doubled if the bridge structure is tuned to make sure the entry of large vehicles onto the bridge deck does not excite destructive natural resonance. The resonance frequency (about

1 to 3 Hz) as a micro-Doppler frequency is imposed on the Doppler frequency shift of the moving tractor trailer.

In [30], a millimeter wave radar operating at 92 GHz is used to analyze micro-Doppler signatures of the jet engine modulation (JEM) lines in an Mi-24 Hind-D helicopter. The research shows that the frequencies of these JEM lines are proportional to the turbine rotation rate and the number of turbine blades. It also shows that only the primary set of turbine blades has contribution to the micro-Doppler signature.

Marple recently proposed a modified time-frequency analysis that provides high resolution without cross-term interference to observe micro-Doppler details induced by extremely weak signal components [31]. The new approach has been applied to returned signals from a two-engine Eurocopter Deutschland BO-105 helicopter illuminated by an X-band radar. With the modified time-frequency analysis, Doppler components due to the forward motion of the fuselage, the micro-Doppler components due to main rotor and tail rotor rotations, and multipath components between the fuselage and rotors can be extracted with at least 70 dB of dynamic range between the strongest and the weakest signal components.

Other new methods for extracting weak micro-Doppler features in the time-frequency domain are still much needed.

References

[1] Mallat, S., and Z. Zhang, "Matching Pursuit with Time-Frequency Dictionaries," *IEEE Trans. Signal Processing*, Vol. 41, No. 12, December 1993, pp. 3397–3415.

[2] Qian, S., and D. Chen, "Signal Representation Using Adaptive Normalized Gaussian Functions," *Signal Processing*, Vol. 36, No. 1, March 1994, pp. 1–11.

[3] Mann, S., and S. Haykin, "The Chirplet Transform: Physical Consideration," *IEEE Trans. on Signal Processing*, Vol. 43, No. 11, 1995, pp. 2745–2761.

[4] Bultan, A., "A Four-Parameter Atomic Decomposition of Chirplets," *IEEE Transactions on Signal Processing*, Vol. 47, No. 3, 1999, pp. 731–745.

[5] Yin, Q., S. Qian, and A. Feng, "A Fast Refinement for Adaptive Gaussian Chirplet Decomposition," *IEEE Trans. on Signal Processing*, Vol. 49, 2001.

[6] Ince, T., K. Leblebicioglu, and G. T. Sayan, "Target Classification by a Time-Frequency Analysis," *1996 IEEE Antennas and Propagation Society International Symposium and URSI Radio Science Meeting Digest*, Baltimore, MD, July 1996.

[7] Kim, K. T., I. S. Choi, and H. T. Kim, "Efficient Radar Target Classification Using Adaptive Joint Time-Frequency Processing," *IEEE Trans. Antennas Propagat.*, Vol. AP-48, December 2000, pp. 1789–1801.

[8]   Dinger, R. J., et al., "Detection of Small Floating Targets on Ocean Surface Using an Ultra-Wideband 150–700 MHz Impulse Radar," *Second International Conference on Ultra-Wideband Short-Pulse Electromagnetics*, Brooklyn, NY, April 1994.

[9]   Halman, J. I., K. A. Shubert, and G. T. Ruck, "SAR Processing of Ground Penetrating Radar Data for Buried UXO Detection: Results from a Surface-Based System," *IEEE Trans. Antennas Propagat.*, Vol. AP-46, July 1998, pp. 1023–1027.

[10]   Xu, X., and R. M. Narayanan, "A Comparative Study of UWB FOPEN Radar Imaging Using Step-Frequency and Random Noise Waveforms," *2000 IEEE Antennas and Propagation Society International Symposium Digest*, Salt Lake City, UT, July 2000, pp. 1956–1959.

[11]   Deng H., and H. Ling, "Clutter Reduction for Synthetic Aperture Radar Imagery Based on Adaptive Wavelet Packet Transform," *Progress in Electromag. Research*, Vol. 29, March 2000, pp. 1–23.

[12]   Chen, V. C., and S. Qian, "Joint Time-Frequency Transform for Radar Range-Doppler Imaging," *IEEE Trans. Aerospace and Electronic Systems*, Vol. 34, No. 2, 1998, pp. 486–499.

[13]   Sparr, T., S. Hamran, and E. Korsbakken, "Estimation and Correction of Complex Target Motion Effects in Inverse Synthetic Aperture Imaging of Aircraft," *Proc. of IEEE Intl. Radar Conference*, 2000, pp. 457–461.

[14]   Wu, Y., and D. C. Munson, "Wide-Angle ISAR Passive Imaging Using Smoothed Pseudo Wigner-Ville Distribution," *Proceedings of IEEE Conference*, 2001, pp. 363–366.

[15]   Stuff, M. A., "Three-Dimensional Analysis of Moving Target Radar Signals: Methods and Implications for ATR and Feature-Aided Tracking," *SPIE Proc. Algorithms for Synthetic Aperture Radar Imagery VI*, Vol. 3721, 1999, pp. 485–496.

[16]   Li, J., and H. Ling, "ISAR Motion Detection and Compensation Using Genetic Algorithms," *SPIE Proc. on Wavelet Applications VIII*, Vol. 4391, 2001, pp. 380–388.

[17]   Barbarossa, S., and A. Farina, "Detection and Imaging of Moving Objects with Synthetic Aperture Radar. Part 2: Joint Time-Frequency Analysis by Wigner-Ville Distribution," *IEE Proceedings—F*, Vol. 139, No. 1, 1992, pp. 89–97.

[18]   Barbarossa, S., "Analysis of Multicomponent LFM Signals by a Combined Wigner-Hough Transform," *IEEE Trans. on Signal Processing*, Vol. 43, No. 6, 1995, pp. 1511–1515.

[19]   Wood, J. C., and D. T. Barry, "Radon Transformation of Time-Frequency Distributions for Analysis of Multicomponent Signals," *IEEE Trans. on Signal Processing*, Vol. 42, No. 11, 1994, pp. 3166–3177.

[20]   Wood, J. C., and D. T. Barry, "Radon Transformation of the Wigner Spectrum," *Proceedings of SPIE on Advanced Architectures, Algorithms on Signal Processing*, Vol. 1770, 1992, pp. 358–375.

[21]   Namias, V., "The Fractional Order Fourier Transform and its Application to Quantum Mechanics," *Journal of the Institute of Mathematics Applications.*, Vol. 25, No. 3, 1980, pp. 241–265.

[22]   Almeida, L. B., "The Fractional Fourier Transform and Time-Frequency Representations," *IEEE Trans. on Signal Processing*, Vol. 42, No. 11, 1994, pp. 3084–3091.

[23]   Zayed, A. I., "On the Relationship Between the Fourier and Fractional Fourier Transforms," *IEEE Signal Processing Letters*, Vol. 3, No. 12, 1996, pp. 310–311.

[24]   Ozaktas, H. M., B. Barshan, and D. Mendlovic, "Convolution and Filtering in Fractional Fourier Domian," *Opt. Rev.*, Vol. 1, 1994, pp. 15–16.

[25]   Lee, S. Y., and H. H. Szu, "Fractional Fourier Transforms, Wavelet Transforms, and Adaptive Neural Networks," *Optical Engineering*, Vol. 33, No. 7, 1994, pp. 2326–2330.

[26]   Mendlovic, D., H. M. Ozaktas, and A. W. Lohmann, "Fractional Correlation," *Appl. Opt.*, Vol. 34, 1995, pp. 303–309.

[27]   Capus, C., Y. Rzhanov, and L. Linnett, "The Analysis of Multiple Linear Chirp Signals," *IEE Seminar on Time-Scale and Time-Frequency Analysis and Applications*, 2000, pp. 4/1–4/7.

[28]   Xia, X. G., et al., "On Generalized-Marginal Time-Frequency Distributions," *IEEE Trans. on Signal Processing*, Vol. 44, No. 11, 1996, pp. 2882–2886.

[29]   Greneker, G., J. Geisheimer, and D. Asbell, "Extraction of Micro-Doppler from Vehicle Targets at X-band Frequency," *Proceedings of SPIE on Radar Technology*, Vol. 4374, 2001.

[30]   Wellman, R. J., and J. L. Silvious, "Doppler Signature Measurements of an Mi-24 Hind-D Helicopter at 92 GHz," ARL-TR-1637, Army Research Laboratory, Adelphi, Maryland, July 1998.

[31]   Marple, S. L., "Large Dynamic Range Time-Frequency Signal Analysis with Application to Helicopter Doppler Radar Data," *ISPA Conference*, 2001.

List of Acronyms

1D	one-dimensional
2D	two-dimensional
3D	three-dimensional
ADS	adaptive spectrogram
AJTF	adaptive joint time-frequency
CAD	computer-aided design
CFAR	constant false alarm rate
CPI	coherent processing interval
CSD	cone-shaped distribution
CWD	Choi-Williams distribution
CWT	continuous wavelet transform
DPCA	displaced phase center antenna
FFT	fast Fourier transform
FOPEN	foliage penetration
FRFT	fractional Fourier transform
GMTI	ground moving target indicator
GPS	global positioning system
GTD	geometrical theory of diffraction
I and Q	in-phase and quadrature-phase
IF	intermediate frequency
INS	inertial navigation system
ISAR	inverse synthetic aperture radar
JEM	jet engine modulation
Joint STARS	Joint Strategic Target Attack Radar System

LFM	linear frequency modulated
LOS	line of sight
MTI	moving target indication
PGA	phase gradient autofocus
PPP	prominent point processing
PRF	pulse repetition frequency
PRI	pulse repetition interval
RF	radio frequency
SAR	synthetic aperture radar
SF	stepped frequency
SNR	signal-to-noise ratio
STAP	space-time adaptive processing
STFT	short-time Fourier transform
TFDS	time-frequency distribution series
VSAR	velocity synthetic aperture radar
WVD	Wigner-Ville distribution

About the Authors

Victor C. Chen received his M.S. and Ph.D. in electrical engineering from Case Western Reserve University, Cleveland, Ohio. Since 1990 he has been with the Radar Division of the U.S. Naval Research Laboratory in Washington, D.C. He is a principal investigator for several research projects. His research interests include synthetic aperture radar, inverse synthetic aperture radar, moving target detection, noncooperative target identification, and time-frequency applications to radar signal and imaging. Before joining the Naval Research Laboratory, he was a senior engineer and staff scientist with Vitro Corporation in Silver Spring, Maryland, Picker International in Cleveland, Ohio, and Technicare Corporation in Solon, Ohio, working on image processing, artificial neural networks, automatic target recognition, moving target detection, 3D motion perception, and MRI and other medical imaging systems. Dr. Chen has published in more than 80 books, journals, and proceedings.

Hao Ling was born in Taichung, Taiwan, on September 26, 1959. He received his B.S. in electrical engineering and physics from the Massachusetts Institute of Technology in 1982 and his M.S. and Ph.D. in electrical engineering from the University of Illinois at Urbana-Champaign in 1983 and 1986, respectively. He joined the faculty of the University of Texas at Austin in September 1986 and is currently a professor in the Department of Electrical and Computer Engineering and holder of the L. B. Mead Professorship in Engineering.

In 1982, he was associated with the IBM Thomas J. Watson Research Center, in Yorktown Heights, New York, where he conducted low temperature experiments in the Josephson Department. He participated in the Summer Visiting Faculty Program in 1987 at the Lawrence Livermore National Laboratory. In 1990, he was an Air Force Summer Fellow at Rome Air Development Center, Hanscom Air Force Base. His principal area of research is computational electromagnetics. During the past decade, he has actively contributed to the development and validation of numerical and asymptotic methods for characterizing the radar cross section from complex targets. His recent research interests also include radar signal processing, fast algorithms for radar image simulation, and automatic target identification.

Dr. Ling has received the National Science Foundation Presidential Young Investigator Award (1987), the NASA Certificate of Appreciation (1991), and several teaching awards from the University of Texas. He is a fellow of the IEEE.

Index

3D target motion, 135–44
  detecting, 140
  model, 138–39, 140
  point-scatterer simulation, 141
  rotational, 139, 143

Adaptive Gaussian extraction, 35–36, 84
Adaptive Gaussian representation, 83–84
Adaptive joint time-frequency (AJTF), 124
  engine, 138
  ISAR image after motion
      compensation, 139
  motion compensation algorithm, 132,
      143
  phase extraction with, 130, 142
  for phase of strongest scatterer, 131
  spectrogram after motion
      compensation, 137
  summary, 135
Adaptive spectrogram (ADS), 35–36, 83
  from conducting strip, 84
  defined, 34
  of test signal, 36, 37
Adaptive time-frequency representation,
      33–36
  applications of, 193–94
  phase estimation with, 128–29
Additive white Gaussian noise, 12
  average noise power, 54
  SNR, 54

Ambiguity function, 13–17
  defined, 13
  of frequency-modulated signal, 14
  for Gaussian pulse, 15
  high value, 13
  for LFM pulse, 16
  peak of, 14
  for rectangular pulse, 15
  of scaled signal, 14
  of SF signal, 17
  in symmetrical form, 13–14
  thumbtack-type, 14
  of time-shifted signal, 14
AN/APY-6 radar, 162–63

Backscattering feature extraction, 194
Bessel functions, 183, 189
Bilinear time-frequency transforms, 36–44
  Cohen's class, 39–42
  TFDS, 42–44
  WVD, 37–39
  See also Time-frequency transforms
Boeing-727 simulation, 132, 133, 135

Cartesian coordinates, 94, 95
Choi-Williams distribution. See CWD
Clutter bandwidth, 150–52
  Doppler bandwidth and, 151
  energy suppression within, 152
  maximum, 152

Coarse range alignment, 125, 127
Cohen's class, 39–42
  defined, 40
  WVD as member of, 40
Coherent processing interval (CPI), 6
Composite translation, 108
Computer-Aided Design (CAD) model, 186
Cone-shaped distribution (CSD), 40–42
  cross-term interference and, 42
  defined, 41
  of test signal, 42
Constant false-alarm rate (CFAR) detection, 47
  block diagram, 60
  illustrated, 58
  in joint time-frequency domain, 57–61
  threshold, 59
  threshold determination, 60
  threshold setup, 47
Continuous wavelet transform (CWT), 32–33
  application to scattering data, 77–79
  basis functions, 33
  defined, 32
  of frequency signal, 77
  multiresolution property, 32
  on inverse Fourier transform, 77
  variable-resolution cells, 33
  See also Linear time-frequency transforms
Cross-range resolution, 22
  defined, 4, 22
  determination, 22
  obtaining, 4–5
  of reconstructed image, 112
CWD, 40–42
  defined, 40, 41
  kernel function, 40
  of test signal, 41

Detection
  CFAR, 47, 57–61
  with GMTI, 162
  of moving targets, 157–65, 196–98
  multiple LFM signal, 196
  signal, 57–58

time-frequency-based SAR image
    formation for, 168–70
  time-frequency transforms for, 48
Displaced phase center antenna (DPCA), 161
  apertures, 162
  defined, 161
  targets, 162
Doppler centroid, 155, 161
Doppler frequency shifts, 98, 99
  instantaneous, 106
  radar/target motions and, 153
  rotational, 99
  from rotational motion, 114
  from rotor blades, 188
  time-varying, 99
  from translational motion, 114
  velocity and, 173
Doppler-frequency trajectories, 128
Doppler parameter estimation, 166–68
Doppler rate
  to be estimated, 160
  filter bank, 160
  of moving targets, 155, 159, 161
  parameter estimation for, 160, 167
  of stationary targets, 155
Doppler resolution, 22
Doppler tracking, 99
  applying, 99
  defined, 99
  in motion compensation, 103
Dwell time, 128
  range profiles vs., 136
  rotation angle relationship, 130

Electromagnetic backscattering, 6–9, 181
Electromagnetic phenomenology, 66–69
ESPRIT, 81
Extraction
  adaptive Gaussian, 35–36, 84
  of dispersive scattering features, 85–89
  in joint time-frequency domain, 61–63
  SNR for, 54–56

False-alarm rate, 59
Fast Fourier transform (FFT), 78, 129
Filter-bank approach, 158–60
Foliage penetration applications, 194

Fourier-based image formation, 21, 102–4
Fourier transform, 25, 101, 178
    of 1D, 133
    of autocorrelation function, 36–37
    fast (FFT), 78
    fractional (FRFT), 196–98
    inverse, 77
    of rectangular pulse, 10
    short-time (STFT), 25–26, 28–31
    of time-shifted rectangular pulse,
        10–11
    of time signal, 26
Fractional Fourier transform (FRFT),
        196–98
    angular parameter and, 196
    defined, 196
    of LFM signal, 198
    robust, 198
    time-frequency transform of, 197
Frequency-domain filtering, 50
Frequency-domain signatures, 173,
        188–90
    illustrated, 188
    information, 190

Gabor spectrogram, 43
Gabor transform, 30, 47, 61, 180
Gaussian window function, 30, 31
Generalized-marginal time-frequency
        distribution, 198
Geometrical theory of diffraction (GTD)
    for canonical conducting structures, 8
    defined, 8
    electromagnetic wave scattering, 67
    as "high-frequency" approximation, 68
    scattering amplitudes and, 68
    theory, 8
Global positioning system (GPS), 135
Goubau mode, 80
Ground moving target indicator (GMTI),
        162–64
    defined, 162
    detection with, 162

Hanning window, 31
Hough transform, 168, 169

Image blurring, 128
Image formation, 195
    Fourier-based, 102–4, 195
    time-frequency-based, 104–6, 195
Inertial navigation system (INS), 135
Inverse Fourier transform, 77
Inverse synthetic aperture radar (ISAR),
        21, 85–89
    adaptive joint time-frequency
        algorithm, 89
    algorithm, 85, 89
    defined, 85
    enhanced images, 87, 89
    images, 85, 86, 87, 88
    joint time-frequency processing, 86
    See also ISAR imaging
ISAR imaging
    after AJTF motion compensation, 139
    challenges, 123
    collection, 124
    formation from measured data, 136,
        137
    generation, 93
    motion compensation in, 123–44
    real-world scenarios, 123
    of simulated Boeing-727, 133, 135
    stationary sensor in, 123
    See also inverse synthetic aperture radar
        (ISAR)

Joint STARS, 162
Joint time-frequency domain, 57–63
    CFAR detection in, 57–61
    signal extraction in, 61–63
    SNR in, 56–57
Joint time-frequency energy distribution,
        102
Joint time-frequency images
    from coated plate, 73
    from conductor-backed periodic
        grating, 74
    from dielectric coated wire, 80
    as distinct features, 71
    from slotted waveguide structure, 76
    via CWT, 79
    via STFT, 79, 80
    via TFDS, 80

Kaiser-Bessel window, 78

Linear frequency modulated (LFM)
        signals, 9
    ambiguity function, 16
    defined, 9
    frequency spectrum, 9, 10
    FRFT of, 198
    with Gaussian envelope, 9
    multiple, detecting, 196
    waveform, 10
Linear phase function, 153
Linear time-frequency transforms, 26–36
    adaptive representation, 33–36
    CWT, 32–33
    STFT, 28–31
    See also Time-frequency transforms

Maclurin series expansion, 152
Maneuvering targets, 107–13
    defined, 107
    dynamics of, 107–8
    imaging with time-frequency-based
        image formation, 108–13
Matched filter, 17–19
    applied to base-band signal, 151
    defined, 17
    output of, 19–20
    properties, 18–19
    response of, 19
Mechanical vibration, 174
Micro-Doppler phenomenon, 173–92
    analysis, 198–99
    defined, 174
    frequency spectrum, 177
    rotation-induced, 181–92
    time-frequency analysis of, 174
    vibration-induced, 174–81
MIG-25 simulation, 109, 110
Millimeter wave radar, 199
Motion compensation, 102–4, 195–96
    blind, 134
    Doppler tracking in, 103
    examples of simulated/measured data,
        131–35
    in ISAR imaging, 123–44
    purpose of, 102
    range tracking in, 103

rotational, 130
standard, 102–3
time-frequency-based, 126–35
Motion compensation algorithms, 124–26
    2D, 140
    AJTF, 132, 137, 143
    goal of, 128
Motion error elimination, 129–31
Moving target indication (MTI), 147
Moving targets, 3, 94–102
    3D, 135–44
    compensation, 102–4
    detection and imaging of, 157–65,
        196–98
    Doppler rate of, 159
    Doppler shift of, 155
    micro-Doppler signatures of, 179–81
    multiple, 113–20
    radar imaging of, 94–102
    radar returns of, 148–54
    rotational motion, 95
    SAR imaging of, 147–70
    time-frequency characteristics, 166
    See also Targets
Multiple-antenna SAR, 161–65
    DPCA, 161–62
    GMTI, 162–64
    velocity SAR, 164–65
    See also Synthetic aperture radar (SAR)
Multiple targets, 113–20
    conventional time-frequency approach
        to, 118
    distinguishing, 115
    Doppler difference between, 114, 115
    geometry of, 116
    identifying, 120
    radar imaging of, 113–20
    resolution analysis, 113–17
    rotational motion, 117
    at same range, 117
    separated, 115
    time-frequency-based image formation
        for, 119–20
    time-frequency-based phase
        compensation for, 117–19
    See also Targets
MUSIC, 81

Parameter estimation, 160
Phase
    alignment, 125
    of baseband signal, 96
    compensation, 118
    detected nonlinearity, 142
    eliminating, 128–29
    estimating, 130
    extracting, 130, 142
    functions, 153
    history function, 118
    interferometric, 163
    of output base-band signal, 151
    returned signal, 114
    shifts, 166–67
    time-derivative of, 96
Phase gradient autofocus (PGA), 126
Point-scatterer model, 66–67
    defined, 66
    electromagnetic scattering theory and,
        67
    illustrated, 66
Polar reformatting, 104, 109
Prominent point processing (PPP)
    model, 124
    multiple algorithm, 126, 129
Prony's method, 81–82
    applying, 81, 82
    defined, 81
    See also Windowed superresolution
        algorithm
Pulse compression, 19–20
Pulse repetition frequency (PRF), 2
Pulse repetition interval (PRI), 9

Quadratic phase coefficient, 129

Radar
    ambiguity function, 13–17
    AN/APY-6, 162–63
    coordinates, 94–95
    defined, 1
    GMTI, 162–64
    line of sight (LOS), 2–3, 7
    millimeter wave, 198
    operational scenario, 2
    range profile, 4, 65–89
    synthetic aperture (SAR), 20–22

two-dimensional image, 4
X-band, 198–99
Radar imaging
    of maneuvering targets, 107–13
    of multiple targets, 113–20
    system based on time-frequency image
        formation, 105
    See also ISAR imaging; SAR imaging
Radar returns, 147–54
    analysis, 152–54
    from CAD model, 187
    clutter bandwidth, 150–52
    phase shifts, 166–67
    range curvature, 149–50
    from rotor blades, 181–84
Radar signatures
    defined, 173
    frequency-domain, 173, 188–90
    micro-Doppler, 179–81
    superposition of, 181
    time-domain, 173, 184–88
    time-frequency, 178, 180, 181,
        190–92
Radial velocity, 155
Range curvature, 149–50
    defined, 150
    illustrated, 150
Range profiles, 4, 178
    aligned, 99
    defined, 21, 65
    dwell time vs., 136
    overlapped, 187, 189
    pulse number vs., 125
    in simple targets, 21
    STFT of, 70
    time-domain signatures as, 187
    time-frequency analysis of, 65–89
    time-frequency representation of,
        70–77
Range resolution, 21–22
Range tracking, 99
    applying, 99
    defined, 98
    in motion compensation, 103
Range walk, 150, 155
Rayleigh distribution, 58, 59
Ray optical descriptions, 67

Ray propagation, 69
Reflectivity density function, 98
Resolution
    cross-range, 4–5
    defined, 4
    Doppler, 22
    range, 21–22
    TFDS, 111
    WVD, 79, 110
Rotational Doppler frequency shift, 97
Rotational motion, 97
    3D, 139, 143
    compensation, 130
    Doppler frequency shift due to, 114
    fast, 109
    multiple targets, 113, 117
    phase compensation and, 118
    polar reformatting and, 104
    removing, 130
    set of, 107
    target, 98
    translational motion combined with,
        119
    See also Translational motion
Rotation-induced micro-Doppler, 181–92
    frequency-domain signatures, 188–90
    rotor blade motion, 181
    rotor blade radar returns, 181–84
    time-domain signatures, 184–88,
        190–92
    See also Micro-Doppler phenomenon
Rotation matrix, 107, 108
Rotor blades
    CAD model, 186
    Doppler frequency shift of, 188
    geometry, 182
    height of, 183
    length, 184
    motion, 181
    radar cross section, 181
    radar returns from, 181–84
    time-domain signatures, 184–85
Round-trip travel time, 1–2

SAR imaging, 147–70
    of simulated Boeing-727, 132
    of stationary targets, 156

system, 100
    target motion effect on, 154, 155–57
    with time-frequency transforms,
        165–70
    See also Synthetic aperture radar (SAR)
Scaling factor, 22
Scattering
    amplitudes, 8, 68–69
    dispersive features extraction, 85–89
    electromagnetic wave, 67
    high-resolution time-frequency
        techniques and, 77–84
Short-time Fourier transform (STFT),
        25–26, 28–31
    basis functions, 30
    defined, 28
    fixed resolution of, 30, 78
    in frequency domain, 29
    with Gaussian window function, 30
    illustrated, 29
    modified, 197
    of range profile, 70
    SNR improvement with, 56–57
    for spectrogram generation, 75, 77
    See also Linear time-frequency
        transforms
Side-looking SAR
    geometry, 148
    range curvature, 149–50
    See also Synthetic aperture radar (SAR)
Signal detection, 57–58
    classical, 57–58
    with fixed threshold, 57
    unknown, 60–61
Signal extraction, 61–63
    in noise, 48–50
    time-frequency expansion/
        reconstruction and, 61–62
    time-frequency masking and, 62
Signal-to-noise ratio (SNR), 2, 48
    additive white Gaussian noise, 54
    average power, 12
    defined, 11
    enhancing, 48
    improvement in time-frequency
        domain, 51–57
    instantaneous power, 12–13, 54

in joint time-frequency domain, 56–57
for signal detection and extraction,
    54–56
with STFT, 56–57
with WVD, 57
Signal waveform, 9–11
Signatures. *See* Radar signatures
Single-aperture antenna SAR, 157–60
    filter-bank approach, 158–60
    parameter estimation, 160
    subaperture focusing, 157–58
    See also Synthetic aperture radar (SAR)
Space-time adaptive processing (STAP),
    157
Spectrogram, 70
    adaptive, 34, 35, 36, 83
    STFT generation of, 75, 77
    of strongest range cell, 133, 136, 137
Stationary targets, 155, 156
    defocused, 157
    SAR imaging of, 156
    time-frequency characteristics, 166
Stepped frequency (SF) signals, 9
    ambiguity function of, 17
    defined, 9
    frequency spectrum, 11
    rectangular pulse, 10
    total bandwidth, 9
    waveform, 11
Subaperture focusing, 157–58
Synthetic aperture radar (SAR), 21–22
    defined, 20–21
    high-resolution map generation, 93
    inverse (ISAR), 21, 85–89
    map, 147
    multiple-antenna, 161–65
    scenes, 147
    side-looking, 148, 149–50
    single-aperture antenna, 157–60
    time-frequency-based image formation
        applied to, 113
    velocity (VSAR), 164–65
    See also SAR imaging
Target coordinates, 95
Target motion
    Doppler frequency shift affected by,
        153

Doppler shift due to, 154
    effect on SAR imaging, 154, 155–57
Targets
    Doppler difference between, 114, 115
    electromagnetic backscattering from,
        6–9
    extended, 113
    initial range, 98
    maneuvering, 107–13
    man-made, 3
    moving, 3, 94–102
    multiple, 113–20
    point, 113
    radial velocity, 155
    reflectivity, 6, 98
    rotational motion, 95, 98
    stationary, 155, 156
    translational motion, 98
Taylor series, 127
Time-domain signatures, 173, 184–88
    as range profiles, 187
    of rotor blades, 184–85
Time-frequency-based image formation,
        104–6
    applied to SAR data, 113
    applying, 104
    effectiveness, 112
    maneuvering target imaging with,
        108–13
    for multiple targets, 119–20
    system illustration, 105
Time-frequency-based motion
        compensation, 126–35
    examples, 131–35
    motion error elimination, 129–31
    phase estimation, 128–29
    presence of 3D target motion, 135–44
Time-frequency coefficients, 61
    extracted waveform, 62
    Gabor, 62, 63
Time-frequency distribution series
        (TFDS), 42–44
    application to scattering data, 79–81
    defined, 42, 43
    resolution, 111
    of test signal, 44
    time-frequency energy, 111

Time-frequency domain
  joint, 56–63
  SNR improvement in, 51–57
  two straight lines in, 169
Time-frequency expansion, 61–62
Time-frequency masking, 62
Time-frequency reconstruction, 61–62
Time-frequency signatures, 178, 180, 181,
      190–92
  blade rotation rate measurement, 192
  distribution, 191
  illustrated, 179, 191
  of vibrating scatterer, 177–79
  See also Radar signatures
Time-frequency transforms, 25–44
  adaptive, 193–94
  bilinear, 36–44
  classes, 26
  for detection/extraction of signals in
      noise, 48
  Doppler parameter estimation with,
      166–68
  of FRFT, 197
  linear, 26–36
  SAR imaging with, 165–70
  trends in, 193–99
Time history series, 101
Time-varying frequency filtering, 48–51
  block diagram, 52
  defined, 50
  illustrated, 49, 50
  iterative, 53
  reconstructed signal and, 49
Translational motion, 97
  Doppler frequency shift due to, 114
  multiple targets, 113
  removing, 130
  rotational motion combined with, 119

target, 98
See also Rotational motion
Velocity SAR (VSAR), 164–65
  defined, 164
  illustrated, 164
  processing, 165
VFY-218 model, 88
Vibration-induced micro-Doppler, 174–81
  signature example of moving targets,
      179–81
  time-frequency signature, 177–79
  See also Micro-Doppler phenomenon
Vibrations, 174
  characteristics of, 174
  modulation due to, 174
  rate of scatterer, 176
  of reflecting surface, 174, 175

Wigner-Ville distribution. *See* WVD
Windowed superresolution algorithm,
      81–83
  Prony's method, 81–83
  time-frequency representation via, 83
WVD, 37–39
  cross-term interference, 39, 110
  defined, 37–38
  frequency marginal condition, 38
  frequency resolution, 110
  group delay property, 39
  instantaneous frequency property, 38
  with linear low-pass filter, 111
  resolution, 79
  SNR improvement with, 57
  of test signal, 40
  for time-frequency-based image
      formation, 110
  time marginal condition, 38
X-band radar, 198–99
Xpatch, 186, 189, 191

Recent Titles in the Artech House Radar Library

David K. Barton, Series Editor

Advanced Techniques for Digital Receivers, Phillip E. Pace

Airborne Pulsed Doppler Radar, Second Edition, Guy V. Morris and Linda Harkness, editors

Bayesian Multiple Target Tracking, Lawrence D. Stone, Carl A. Barlow, and Thomas L. Corwin

Computer Simulation of Aerial Target Radar Scattering, Recognition, Detection, and Tracking, Yakov D. Shirman, editor

Design and Analysis of Modern Tracking Systems, Samuel Blackman and Robert Popoli

Digital Techniques for Wideband Receivers, Second Edition, James Tsui

Electronic Intelligence: The Analysis of Radar Signals, Second Edition, Richard G. Wiley

Electronic Warfare in the Information Age, D. Curtis Schleher

EW 101: A First Course in Electronic Warfare, David Adamy

Fundamentals of Electronic Warfare, Sergei A. Vakin, Lev N. Shustov, and Robert H. Dunwell

Handbook of Computer Simulation in Radio Engineering, Communications, and Radar, Sergey A. Leonov and Alexander I. Leonov

High-Resolution Radar, Second Edition, Donald R. Wehner

Introduction to Electronic Defense Systems, Second Edition, Filippo Neri

Introduction to Electronic Warfare, D. Curtis Schleher

Microwave Radar: Imaging and Advanced Concepts, Roger J. Sullivan

Modern Radar System Analysis, David K. Barton

Printed in the United States
84497LV00002B/274-297/A